I0132278

INTERNATIONAL DEVELOPMENT IN FOCUS

Jobs from Agriculture in Afghanistan

Izabela Leao, Mansur Ahmed, and Anuja Kar

WORLD BANK GROUP

Contents

Foreword

Agriculture has always been the primary pursuit of the people in Afghanistan—around 70 percent of the population live and work in rural areas, mostly on farms. Of off-farm employment, including in urban and peri-urban areas, a large share of employment is in agriculture-related sectors and food processing, and the agricultural industry accounts for most of the exports and about 40 percent of manufacturing. Inherently, the Afghan economy is centered on the agriculture sector.

Given the numerous challenges faced by the sector in the Afghan context, however, agriculture has not yet been able to realize its full potential. Revitalizing agriculture and creating agriculture jobs is thus a priority for the government of Afghanistan, as the sector can play an important role in reducing poverty and sustaining inclusive growth in the country.

I am pleased to note that the findings from the *Jobs from Agriculture in Afghanistan* report, which has been collaboratively prepared by the World Bank and the government of Afghanistan, reinforces the message that, if necessary policy decisions are made and an enabling environment is created, the agriculture sector can play an even more prominent role in poverty alleviation, job creation, improvement of livelihoods, and achievement of food security and nutrition for the rural population.

We acknowledge the significant contributions made by Afghan women in the various stages of agriculture, including cultivation of animals, and the planting and harvesting of crops. There is little doubt that women play a crucial role in not only enhancing food security and nutrition, but also in boosting the economy and guaranteeing sustainable development and rural stability.

This is an important and timely publication, as a renewed momentum and more focused interventions are underway for ensuring that we create sustainable jobs for Afghans, especially youth and women. The report's findings will inform the policy discourse and enrich the choices available for the government of Afghanistan, and its partners, to embark on a set of structured interventions that will lead to full utilization and maximum realization of the energy and potential that the agriculture sector presents.

I would like to acknowledge the great amount of effort placed in preparing and finalizing this report, and would like to thank the task team leader and the team for their excellent and tireless work to present such a report, which fills important gaps in data analyses on jobs, viewed through an agriculture lens, and which provides clear recommendations for the Ministry of Agriculture, Irrigation, and Livestock and its partners to effectively implement job creation policies in the agriculture sector.

Nasir Ahmad Durrani
Minister
Ministry of Agriculture, Irrigation, and Livestock
Afghanistan

Foreword

Agriculture is a pillar of development and national security in Afghanistan. The sector employs 40 percent of the total labor force, and more than half of the rural workforce is involved in agriculture.

The *Jobs from Agriculture in Afghanistan* report is the result of a partnership between the government of Afghanistan and the World Bank. It extensively assesses agricultural employment support and creation in Afghanistan to guide policy makers in enhancing agriculture's contribution to jobs in the years ahead.

It analyzes three aspects of jobs in the sector. First, it evaluates the current jobs structure in rural areas and finds that rural jobs are concentrated in cereal agriculture, especially in wheat, reflective of the differences in relative returns to activities within agriculture. Second, it analyzes the inclusive nature of agriculture jobs for vulnerable groups, such as women, youth, the landless, and the bottom 40 percent of income earners. The analysis finds that although agriculture jobs are inclusive, many women and youth participate as voluntary family workers, since they are unable to access markets and/or find paid jobs in the non-farm sector. Third, the report evaluates the role of public and private sector interventions in supporting job creation in agriculture. The analysis finds that these interventions can work and that there is significant scope to scale them up.

As development proceeds in Afghanistan, as elsewhere, the share of people employed directly in farming will decline with an increase in the share employed in the broader food system, such as in agriculture-related manufacturing and services. For the near future, as the economy adjusts to changes brought about by the decline of aid flows and foreign military expenditures, raising the returns to activities within agriculture will be crucial to absorb new farm-level workers, as well as to raise the incomes of those already engaged in agriculture. Over time, growth in the broader food system, including in food storage, processing, distribution, transportation, retailing, and preparation, should be promoted, as it can make a major contribution to overall growth in job opportunities for men and women.

The findings in this report offer a guide to the formulation of policies that can enhance agriculture's contribution to jobs in Afghanistan, an aspect that is key for overall poverty reduction in the country.

Juergen Voegele
Senior Director
Agriculture Global Practice
The World Bank

Shubham Chaudhuri
Country Director
Afghanistan
The World Bank

Acknowledgments

At the request and with the support of the Ministry of Agriculture, Irrigation, and Livestock (MAIL), and with the collaboration of the Ministry of Rehabilitation and Rural Development (MRRD) of the government of Afghanistan, this report was a multisectoral collaboration among the World Bank's Global Practices of Agriculture; Poverty; Social, Urban, Rural, and Resilience; Water; Trade and Competitiveness; Education; Social Protection and Labor; and the International Finance Corporation.

The report was prepared by Izabela Leao (Task Team Leader), under the overall guidance of Shobha Shetty (Practice Manager), with contributions from a core team including Mansur Ahmed and Anuja Kar, and with the support of Parmesh Shah, Dorte Verner, and Madhur Gautam. Major contributions, through research, case studies, or as inputs to the study, were provided by a multisectoral team that included: Amanullah Alamzai, Azada Hussaini, Abhishek Saurav, Camilla Schloss, Christina Wieser, Hazem Ibrahim Hanbal, Jana El-Horr, Lida Homa, Mabruk Kabir, Mir Ahmad Ahmad, Mohammad Yasin Noori, Ramziath Adjao, Rohan Selvaratnam, Shubha Chakravarty, Yehia Khedr Eldozar, and Yuxuan Zhao. Abhishek Saurav and Yehia Eldozar were lead contributors to the Omaid Bahar Fruit Processing case study. Spotlight box 3.1 was authored by Mohammad Yasin Noori and Shankar Narayanan; spotlight box 3.2 was authored by Jana El-Horr and Shubha Chakravarty; and spotlight box 3.3 was authored by Camilla Schloss and Tazeen Hasan.

Izabela Leao, Mansur Ahmed, and Anuja Kar authored the chapters.

The report benefitted from the guidance of H.E. Nasir Ahmad Durrani, Minister of MAIL, as well as in his former capacity as Minister of MRRD. It also benefitted from the guidance and support of MAIL's Former Minister, H.E. Assadullah Zamir; Deputy Minister of Finance, H.E. Abdul Qadeer Jawad; and the Strategic Planning Advisor, Shakir Majeedi. Collaboration with MRRD's Director for the Afghanistan Rural Enterprise Development Project (AREDP), Rahmatullah Quraishi, was also beneficial.

The authors also benefitted from inputs from the Donor Community Working Group for Agriculture in Afghanistan during a consultation meeting held in Kabul in May 2016. Participants included Haroon Khawar and Habibur Rahman (Japan International Cooperation Agency); McDonald C. Homer (United States Agency for International Development); Byron Syler and Muhebullah Latifi

(Food and Agriculture Organization of the United Nations); Noor Hakimyar and Ismail Qarizada (U.K. Department for International Development); Frank Mussugnug and Frank O'Sullivan (German Society for International Cooperation); Simon Puckett and Zamarai Samin (Australian Embassy); and, Gael Griette, Giampiero Muci, Filippo Saracco, and Simone Raudino (European Union).

The team is particularly thankful to members of the World Bank projects that were analyzed, both from the World Bank and from government, who supported us with data collection, guidance, consultations, and inputs. The authors thank Sayed Usman Safi, Abdul Fulady Wahab, Sayed Milad Waizi, Shaima Ahadi, Amanullah Alamzai, and Hazem Ibrahim Hanbal (National Horticulture and Livestock Project); Pervaiz Ahmad Naseri, Saifullah Sahibzada, Ahmad Haseeb Payab, Baryalay Baz, Toru Konishi, Mir Ahmad Ahmad, and Bayarsaikhan Tumurdavaa (On-Farm Water Management Project); Rahmatullah Quraishi, Md. Salim Mastoor, Kamran Akbar, Winston Dawes, and Azada Hussaini (AREDP); Rasoul Rasuli, Khyber Farahi, Abdul Rahman, Ahmad Mukhtar Sabri, Nasrullah Ahmadzai, Abdul Saboor Mohammad Ajan, Brigitta Bode, Mir Siamuddin Abedi, Phillippe J. de Naurois, Md. Ateeq Zaki, and Naila Ahmed (National Solidarity Program III). The team is also grateful to Mustafa Siddiqi, CEO and President of Omaid Bahar Fruit Processing Limited in Kabul.

The team also received inputs and support from other colleagues, including William Magrath, Grahame Dixie, Eija Pehu, Ed Keturakis, Matthew Morton, Najla Sabri, Ladisy Chengula, Silvia Redaelli, Mary Hallward-Driemeier, Luc Christiaensen, Nathalie Lahire, Sanna Liisa Taivalmaa, Sandra Broka, Pushina Kunda Ng'andwe, Melissa Williams, Edna Massay Kallon, Vivek Prasad, Jamie Greenawalt, Karishma Wasti, Sudha Bala Krishnan, Beibei Emily Yan, Teresa A. Peterburs, Ding Xu, Flore Martinant de Preneuf, Wael Zakout, Markus Kostner, Natasha Hayward, and Binayak Sen.

Administrative and team support was provided by John Prakash Badda, Gizella Diaz Munoz, Venkat Ramachandran, Nilofar Amini, Parwana Wawreena Nasiri, Folad Hashimi, Mirwais Farooq Mujaddidi, Zabiullah Ahrary, Hamidullah Safi, Mohammad Asif Qurishi, Beaulah C. Noble, Najibullah Ziar, Sayed Ahmad Shah Hashimi, Saboor Fazai, Abdul Saboor, Fada Amir, Abdul Manan, Haroon, Attiqullah and Nasratullah. External communications support was provided by Raouf Zia, Yann Doignon, Rafi Mohammad Farooq, and Joe Qian.

The study was peer reviewed by Gladys Lopez-Acevedo (Lead Economist, Chief Economist's Office for the South Asia Region), Wagma M. Karokhail (Country Officer for Afghanistan, International Finance Corporation), and Parmesh Shah (Global Lead for Livelihoods and Jobs in the Agriculture Global Practice). Garrison Spik edited the report.

The study benefitted from the overall guidance of Shubham Chaudhuri (Country Director for Afghanistan, SACKB), Robert J. Saum (Former Country Director for Afghanistan, SACKB), Abdoulaye Seck (Manager of Operations for Afghanistan, SACKB), Stephen N. Ndegwa (Former Manager of Operations for Afghanistan, SACKB), Wezi Marianne Msisha (Sr. Operations Officer, SACKB), Fei Deng (Country Program Coordinator, SACKB), Marcia Whiskey (Sr. Country Program Assistant, SACKB), and Macmillan Anyanwu (Former Sr. Country Officer for Afghanistan, SACKB).

The team benefitted from the overall and continued support of the Agriculture Global Practice senior management, including Juergen Voegele, Ethel Sennhauser, Shobha Shetty, Simeon K. Ehui, Martien van Nieuwkoop, Preeti Ahuja, and Rob Townsend, as well as from the support of Martin Rama, Chief Economist for South Asia.

Last, the team is grateful to the World Bank Group Country Office in Afghanistan, and, most important, to the Afghanistan country authorities for their support and trust throughout this endeavor.

About the Authors

Izabela Leao is a Rural Development Specialist in the Agriculture Global Practice of the World Bank, where she works on analytical and operational tasks in the areas of jobs and agriculture, rural economy, youth employment, innovation and entrepreneurship, and conflict/fragility. She joined the World Bank as a Young Professional in 2013 and has since been focusing her work on Afghanistan, Bhutan, and India. She has also worked on the Arab Republic of Egypt, Belarus, Côte d'Ivoire, Nepal, Pakistan, Tajikistan, and Myanmar. Prior to joining the World Bank Group, she spent five years at the United Nations, working on programs in the prevention of transnational organized crime and on programs of disarmament, demobilization and reintegration—focusing on economic reintegration of young ex-combatants, with a particular focus on skills training and employment creation in Colombia, Nepal, and Sierra Leone. Additionally, in 2011, she cofounded a nonprofit organization in Sierra Leone focused on improving the lives of marginalized youth through educational and life skills training. Leao holds a master's degree in history and political science from Pittsburgh State University, focusing on social protection policy for victims of transnational human trafficking, and a doctorate in political science from University of Milan, Italy, focusing on youth agency, state collapse, and unemployment in Sierra Leone.

Mansur Ahmed is an Agriculture Economist in the World Bank's Agriculture Global Practice, where he works on analytical and operational tasks in the areas of agriculture and food security, rural nonfarm economy, and rural jobs. His research interests include econometric and statistical modeling, productivity and efficiency analysis, poverty and social impact analysis, impact evaluation, and inclusive jobs in the context of developing countries. Over the past few years, he has worked extensively in the areas of agricultural productivity, rural jobs, and rural nonfarm economy in Afghanistan, Bangladesh, China, India, Mali, and Myanmar. Prior to joining the World Bank, he worked at the Bangladesh Institute of Development Studies and at the South Asia Network on Economic Modeling in Bangladesh. He holds bachelor's and master's degrees in economics from the University of Dhaka, and a doctorate in economics from North Carolina State University.

Anuja Kar joined the World Bank's Agriculture Global Practice as an Economist in June 2015. She is currently engaged with the Global Agriculture and Food Security Program as an Economist/Monitoring and Evaluation Specialist. Prior to joining the Agriculture Global Practice, Kar spent two years (also within the Bank) building econometric models using household- and farm-level surveys for Bangladesh, India, and Sri Lanka. Her key areas of focus are food security, agriculture productivity, labor markets, and foreign direct investment. Before joining the development sector, she worked in a private institution for two years, where her primary responsibilities included covering the Philippines' macroeconomics and developing macroeconometric models and tools for emerging markets. She holds a master's degree with distinction in public policy from the University of Nottingham, United Kingdom, specializing in food security in Asia, and a master's degree in economics from the University of Calcutta, with majors in advanced economic theory (honors) and econometrics.

Abbreviations

Af	Afghan Afghani (local currency)
ALCS	Afghanistan Living Condition Survey
ARAZI	Afghanistan Independent Land Authority
AREDP	Afghanistan Rural Enterprise Development Program
ARTF	Afghanistan Reconstruction Trust Fund initiative
ASR	Agriculture sector review (World Bank)
CDC	community development councils
CDD	community-driven development
CPF	Country Partnership Framework
CSO	Central Statistics Organization
DFID	U.K. Department for International Development
FTE	full-time equivalent
GDP	gross domestic product
IDPs	internally displaced persons
ISR	implementation status and results
LFPR	labor force participation rate
MAIL	Ministry of Agriculture, Irrigation, and Livestock
MCG	Maintenance Cash Grant Project
MRRD	Ministry of Rural Rehabilitation and Development
NHLP	National Horticulture and Livestock Project
NRVA	National Risk and Vulnerability Assessment
NSP	National Solidarity Program
OFWM	On-Farm Water Management Project
SMEs	small and medium-sized enterprises
TFP	total factor productivity
USAID	U.S. Agency for International Development

All dollar amounts are U.S. dollars unless otherwise indicated.

Executive Summary

About 70 percent of the population in Afghanistan lives and works in rural areas, mostly on farms. Therefore, an analysis through an "agricultural jobs lens" offers insights into the state of the rural labor market, and can facilitate the formulation of effective job creation policies for the rural population. To these ends, the report explores agriculture's direct and indirect roles in recent rural employment and income dynamics using the National Risk and Vulnerability Assessment Survey 2011–12 and the Afghanistan Living Condition Survey 2013–14. It concludes by identifying the key development interventions and areas of policy formulation that could further the creation of more, sustainable, and inclusive employment opportunities for rural Afghans.

SUMMARY MESSAGES

Afghanistan's rural economy is experiencing an influx of youth workers into the labor force, increasing competition for every new job. Although this new generation has more competitive human capital, the rural economy is not yet equipped to absorb all workers into the labor market. As a result, more than 50 percent of rural youth workers are involved in agriculture and livestock, mostly as unpaid family workers.

At 29 percent, the female labor force participation rate in rural areas continues to be low and 60 percent of employed women work in the livestock sector. Four of every five female rural workers are unpaid family workers, compared with only one of every five male workers. Furthermore, around 20 percent of employed female workers are involved in the food and handicraft manufacturing and processing sectors; the figure is only about 2 percent for male workers.

The low share of agricultural income, despite high agricultural employment in the agriculture and livestock sectors, is primarily driven by the limited market participation and the high number of unpaid family workers. Few of the rural households that own garden plots participate in the market or earn income from orchards. Similarly, market participation is low among rural households that raise livestock. Moreover, youth and women constitute a large portion of this unpaid workforce.

1

Although diversification toward high-value agriculture can support gainful job creation, the crop agriculture subsector is not diversified and is overly concentrated on wheat. Lack of diversification has made farm households vulnerable to stagnant or declining wheat prices in local markets. While farmers continue to produce wheat and other food crops for subsistence to ensure food security and nutrition for their families, lack of profitability in wheat production may prompt them to cultivate poppy on irrigated land. (This helps explain a gradual rise of areas under poppy cultivation). Moreover, wheat-based agriculture does not create gainful employment, particularly for women and youth.

If garden plot owners and livestock growers are provided with technical and financial support and greater access to market facilities, horticulture and livestock have great potential to further the creation of more, sustainable and inclusive jobs. Commercial production of fruits and nuts in garden plots, as well as livestock products such as meat, milk, and dairy, would not only increase garden owners' and livestock growers' income and employment—it would help create new jobs for young workers across the fruits, vegetables, nuts, meat, milk, and dairy value chains. By improving the horticulture and livestock economy, the government could also increase the employment share of the food processing sector.

POLICY RECOMMENDATIONS

The key challenge for policymakers and development practitioners is to create more jobs, including better-skilled and more inclusive jobs for youth and under-employed workers, including women. With financial support and increased access to market facilities, horticulture and livestock have great potential for job creation in rural Afghanistan. Policymakers need to address restricted credit and market access, and the weak rural infrastructure (connectivity, transportation, communications, power, and water) to create more efficient supply chains. The following policy recommendations will support more, sustainable, and inclusive job growth through agriculture and rural development.

Diversification toward high-value crop agriculture and livestock. While policies to improve crop productivity, especially wheat, should be in place, policies to diversify agriculture toward high-value agriculture including fruits, vegetables, and livestock should be prioritized. Expansion of irrigation facilities and improved seeds availability can support productivity growth in crop agriculture and reduce underemployment among subsistence farmers. Bringing new areas under irrigation can generate more jobs in rural areas. Horticulture and livestock also have great potential for sustainable and inclusive job creation.

Linking farmers to markets through continued investment in connectivity and infrastructure. Development of agricultural value chains is key to raising productivity and supporting job creation in agriculture. Continued investments in rural roads and other local infrastructure, information and communication technology, and reliable and affordable access to energy are necessary to enable local producers of crops and horticulture and livestock products to access markets and increase agricultural productivity. Rural infrastructure and improved rural-urban connectivity are crucial for the development of national value chains for agricultural products. Policies and investments to improve women's access to markets are also important to catalyze the livestock and horticulture subsectors, and manufacturing and processing sectors, where female workers are predominantly employed.

A balanced development strategy for an enabling environment for farm and nonfarm sectors. There is strong evidence that rural sectors are interdependent; therefore, both the farm and nonfarm sectors must be targeted for sustainable inclusive growth and employment creation. Increased agricultural productivity can boost demand for nonfarm services and products, and a vibrant nonfarm sector can increase demand for high-value agricultural products. Thus, the sectors support each other, raising productivity and generating more, sustainable, and inclusive jobs in rural areas. To operationalize this balanced development strategy, operations in the agriculture sector can be developed to strengthen forward, backward, and consumption linkages, providing opportunities to establish value chains that, if exploited adequately, can support economic growth in the on-, off-, and nonfarm economies.

Access to finance and providing skills development training for job creation in the nonfarm sector. Access to finance and provision for skills development training must also be prioritized, particularly for women and youth. The analysis shows that literacy supports women to join the workforce, and evidence from agricultural and rural development interventions shows that women who have access to finance and linkages to markets are successfully engaging in nonfarm activities and improving their livelihoods. Policymakers and donors need to stress policies and interventions that ease financial constraints and improve the skills of the rural workforce, mainly for the most vulnerable groups, to generate more, sustainable, and inclusive jobs.

Strengthening the private sector presence in agriculture and its linkage with the public sector by means of promoting agribusiness. Private sector efforts should be underpinned by macro institutional, regulatory and business environment support to realize the potential of agriculture. This study shows that two policy levers can enhance the growth potential of jobs in the agro-processing sector. First, enhanced provisions of investments and advisory services to promising agro-processing firms are critical for strong job creation. This type of growth can create wage-bearing jobs for local economies, as well as in the regions from which inputs are sourced and where products are distributed and sold. Second, government policy must support the increased use of vertical integration (or at least coordinated linkages) to mitigate risks in the supply chain. Interventions to improve cross-sectoral linkages in the supply chain may offer agro-processing firms of all sizes better prospects to exploit market opportunities through flexible business models and lower capital requirements.

POLICY FOCUS ON IMPROVING THE DESIGN STRUCTURE OF JOBS MEASUREMENT IN AGRICULTURE AND RURAL DEVELOPMENT

A strong private sector, with public policy support, can support job creation across agricultural value chains and improve the quality of existing jobs, especially for youth and women. The agriculture supply chain is a network of resources and materials that flows from the origin (farm level) to customers; each section plays a role in satisfying market demand downstream and leveraging margin opportunities by transmitting materials, intermediary products and services, and final goods. Importantly, each section provides economic opportunities, and the network supports permanent and temporary jobs that offer income opportunities. For example, this study found that over the past five years, for every 10 jobs created in a lead firm in the fruit processing sector, the upstream and downstream network supported an average of eight jobs.

From the public sector side, to fully realize agriculture's potential to create more, sustainable, and inclusive jobs, it is necessary to design and implement projects with a stronger, clearer jobs focus. To date, jobs results have often been mere by-products of agriculture and rural development operations. Explicitly considering the jobs challenge in the ex-ante project design and results framework will poise the agriculture portfolio to expand its impact on the multidimensional jobs agenda. Instead of a combination of complementary projects to achieve a sustainable impact, Afghan farmers need to have the necessary agricultural skills, marketing and trading knowledge, the required access to transport and markets, and a favorable macroeconomic environment so they can use available resources more effectively. This will also help to create sustainable and inclusive jobs at higher levels of value chains.

From the private sector side, designing effective policies for job creation in agriculture requires analytical rigor and advances in measuring the effects of proposed interventions. Policy planning in Afghanistan can be challenging due to a lack of analytical evidence around the effectiveness of private sector interventions and their impact on employment, incomes, and growth. Improving the availability of administrative data and statistical assets can lower this information barrier and aid the design of interventions. Measuring induced effects requires a lot of information, including data on employee incomes, household savings and expenditures, and geographical consumption patterns. These data need to be synthesized to compute increased revenues for businesses, which can then be used to calculate the number of resultant jobs. Most of the data is administrative in nature; national statistical agencies and affiliated government departments can make this data available.

An intensive policy discourse to create more, sustainable, and inclusive jobs in rural areas should be channeled through promoting farm, nonfarm, and off-farm linkages. The backward linkages with input suppliers (such as seed, feed and fertilizer, pest and disease identification and control, family farms, aggregators, and cooperatives) and intermediate service providers (such as transporters and packing sheds) further create jobs and income opportunities locally and in other regions. Similarly, through forward linkages with distributors, wholesalers, and retailers, agro-processors contribute to additional job creation and economic spillovers. Direct and indirect job creation is further complemented by income-type effects that result from rising incomes and expenditures on consumer goods and services, which require strong linkages among activities in the public and private sectors.

1 Jobs and Agriculture Nexus in Fragile Contexts

THE AFGHANISTAN CASE

INTRODUCTION

The government of Afghanistan and the World Bank Group recognize that revitalizing the agriculture sector will play an important role in poverty reduction and sustained growth, primarily through job creation, improved productivity, and inclusiveness. Much of the agricultural infrastructure was destroyed during three decades of war. In the 1970s, before the war, Afghanistan was a lead international supplier of horticultural products, with export earnings from the sector accounting for 48 percent of total export revenues (Yousufi 2016, 36–42). Until the late 1970s, Afghanistan was the world's largest raisin producer, supplying 20 percent of global exports (World Bank 2011). According to 2008–09 data, that share has fallen to 2.3 percent, and agricultural production is half of its pre-war level (World Bank 2013). Moreover, Afghanistan used to be self-sufficient in meat, milk, and cereals, and was a significant exporter of wool, carpets, and leather goods. Despite this continued decline, the economy remains largely agrarian, and agriculture is the second largest sector after services.

The World Bank's 2014 Agriculture Sector Review (ASR), *Revitalizing Agriculture for Economic Growth, Job Creation and Food Security,* argued that driving agricultural expansion would require focusing on subsectors with significant catch-up potential in the short term to raise productivity, such as irrigated wheat, horticultural crops, and intensive livestock production.[1] In 2014, these subsectors accounted for 66 percent of agricultural gross domestic product (GDP) and 36 percent of total agricultural employment. Based on ASR estimates, with the right mix of policies and investments, these subsectors could more than double agricultural GDP over the next 10 years, implying an addition of 1.3 million full-time equivalent (FTE) jobs—those with 200 work days per year—within the next decade. Yet, it suggests that as in the Green Revolution model, the government of Afghanistan would need to play a lead role in coordinating the strategy in value chains to encourage agricultural growth and in overcoming constraints, for example, in providing access to finance, improving poor on-farm water management and correcting insufficient irrigation infrastructure (World Bank 2014a, ix–16).

The ASR also stated that agriculture could make a significant contribution to employment through expansion of high-value horticulture and livestock production, together with broad rural poverty reduction interventions based on community empowerment and small-scale public works, and by supporting the development of small and medium-sized enterprises. According to the ASR, livestock contributes 15 percent of agricultural GDP ($680 million annually) and creates 1.1 million FTE jobs (of which 15 percent are off-farm); horticulture contributes 34 percent of agricultural GDP ($1.4 billion annually) and involves more than 2 million people (World Bank 2015a, 10–11). Agricultural growth could generate 3.2–3.4 million FTE jobs, including backward and forward linkages. Based on ASR estimates, irrigating arable land for crop production could create 33–60 percent more jobs per hectare than rain-fed farming, and shifting from wheat to production of some horticultural crops could triple—even quadruple—the labor input per hectare. Likewise, jobs in agriculture could be further stimulated by supporting access to credit, land, and markets, promoting intensive livestock production and high-value crops, and creating economic opportunities for women along selected agricultural value chains (World Bank 2014a, 8). The positive spillover effects from the revitalization of the agriculture sector include the potential for countering illicit opium poppy production, boosting agro-based industry growth, and strengthening the role of women in agriculture through decision-making and a livelihoods approach, both key for job growth (World Bank 2012, 2014a).

STUDY CONTEXT AND BACKGROUND

Afghanistan is a fragile country that has seen almost constant conflict since the Soviet invasion in 1979. Although it made significant economic and social progress (from a very low base) in the 2002–12 pre-transition period, socioeconomic outcomes have mostly remained poor. In effect, poverty levels have been increasing since 2012, with 39 percent of the population living below the national poverty line between 2012 and 2014. Regional disparities have driven stagnant poverty rates, with rural areas lagging, particularly in the northeastern region (World Bank 2016b). Moreover, poverty is strongly associated with deprivation in education, employment opportunities, and access to basic services; the increasing inequality accounts for the lack of poverty reduction (figure 1.1) (World Bank 2016a, 10). Last, the unprecedented political, security, and economic transition since 2012 has led to a visible decline in economic performance, threatening the foundations of stability and progress.

The poor in Afghanistan tend to live in rural areas of the country's 34 provinces, deriving their livelihoods from agriculture, and are prone to being underemployed or employed in casual and vulnerable jobs. To understand the spatial distribution of income in rural areas, we plotted per capita income and its growth for rural people (map 1.1). We observed that people in the rural areas neighboring Kabul are the most affluent, and that per capita income is higher in the provinces of Kabul and Logar (Af 25,000+). Per capita income in Helmand, Khost, Nimroz, Paktya, and Panjshir provinces is also in the high-income range (Af 20,000–25,000). Per capita income is generally low in the west central regions, with Daykundi, Farah, Ghor, and Uruzgan about one-quarter less than in Kabul and Logar. The left panel in map 1.1 shows that provinces in the east and south experienced their highest income growth between 2012 and 2014,

FIGURE 1.1

Poverty and inequality in Afghanistan

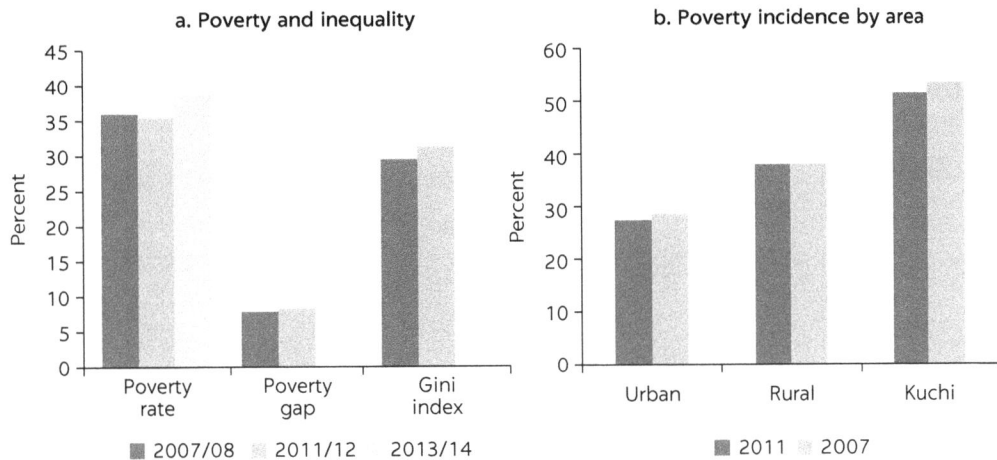

a. Poverty and inequality

b. Poverty incidence by area

2007/08   2011/12   2013/14

2011   2007

Source: Afghanistan Poverty Status Update 2015, The World Bank.

MAP 1.1

Per capita income and its growth in rural Afghanistan, 2012–14

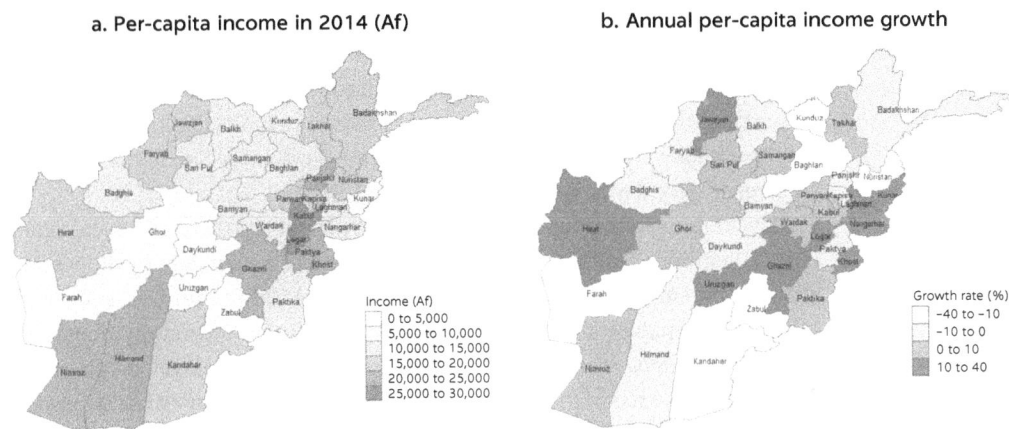

a. Per-capita income in 2014 (Af)

b. Annual per-capita income growth

Sources: Based on NRVA 2011–12 and ALCS 2013–14.

while about half of all provinces experienced income reduction in real prices during the same period. The map also shows that rural households in six provinces had extensive income reduction (more than 10 percent), while the provinces that showed positive income growth were in and around Kabul. The country's urbanization rate is low, and about half of the total urban population lives in Kabul (Central Statistics Organization [CSO] 2016).

According to the World Bank Afghanistan's Systematic Country Diagnostic (SCD), reducing poverty will require strong employment generation (more sustainable, less vulnerable), with special attention to rural areas. Yet, while the economy has been shifting from agriculture to services, primarily because of official development assistance (foreign aid), not private sector investments, the challenge of structural transformation has persisted. In 2012, the services sector accounted for more than half of GDP, while the share of agriculture (excluding opium[2]) had declined to about 25 percent of GDP (figure 1.2).

FIGURE 1.2

Sectoral dynamics of employment and value-added in Afghanistan

a. Total employment, by
sector of economic activity

b. Share of value-added

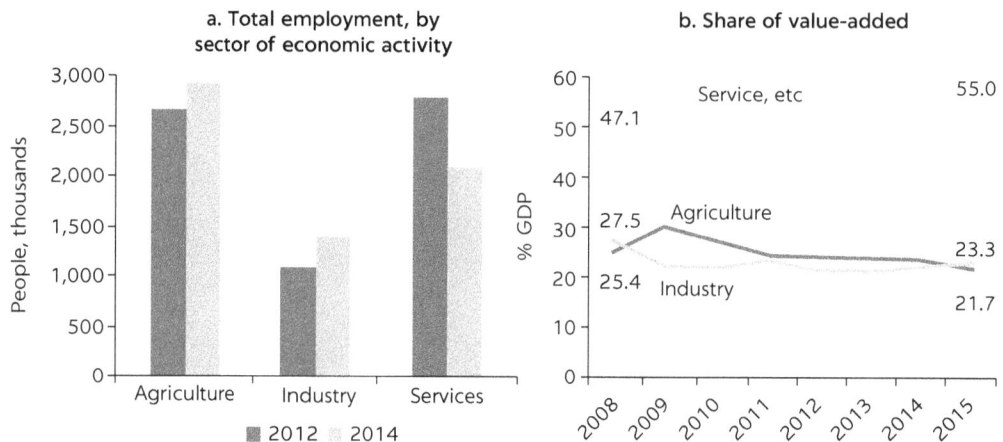

Sources: Based on NRVA 2007–08 and 2011–12, and ALCS 2013–14.

This structural shift is a common trend in developing countries; in Afghanistan, however, it differs because it has been driven largely by an inflow of international aid. As a result, the expansion of non-agriculture employment opportunities has not reduced poverty, although aid-led growth contributed to the creation of new non-agriculture jobs and the reduction of more vulnerable employment: Eighty percent of jobs created in the services sector were informal, and more than 50 percent were day labor jobs (World Bank 2015b).

The Afghanistan Country Partnership Framework (CPF) for fiscal years 2016–20 emphasizes the role of agriculture sector development in supporting growth and employment. The pathway to job creation requires a concerted effort to link public and private sector interventions. This study identified numerous challenges and opportunities, paving the way for developing a policy discourse for the sector to achieve goals in the CPF. The primary goals go beyond simply creating more jobs, to creating sustainable and inclusive jobs. Evidence from World Bank interventions in agriculture and rural development projects in Afghanistan suggests that more, sustainable, and inclusive jobs can be supported by the development of community-based enterprises; integrated value chains in rural areas; improved access to services and resources via non-governmental organizations (NGOs) and government agencies; improved technologies in livestock, high-value crops, and orchards; and efficient water use.

AGRICULTURE AND JOB CREATION IN THE CONTEXT OF FRAGILITY IN AFGHANISTAN

Historically, agriculture has dominated Afghanistan's economy and has been one of the main contributors to economic growth. Since 2000, however, the economy has averaged 10 percent annual growth, with most GDP growth originating in the nonfarm sector, especially the service sector. (The rise of foreign aid for reconstruction and security has supported service-led growth since the fall of Taliban in 2001.) A near-stagnant agricultural GDP in recent years has led to a

FIGURE 1.3
Gross domestic product (GDP), GDP growth, agriculture GDP, and share of agriculture in total GDP

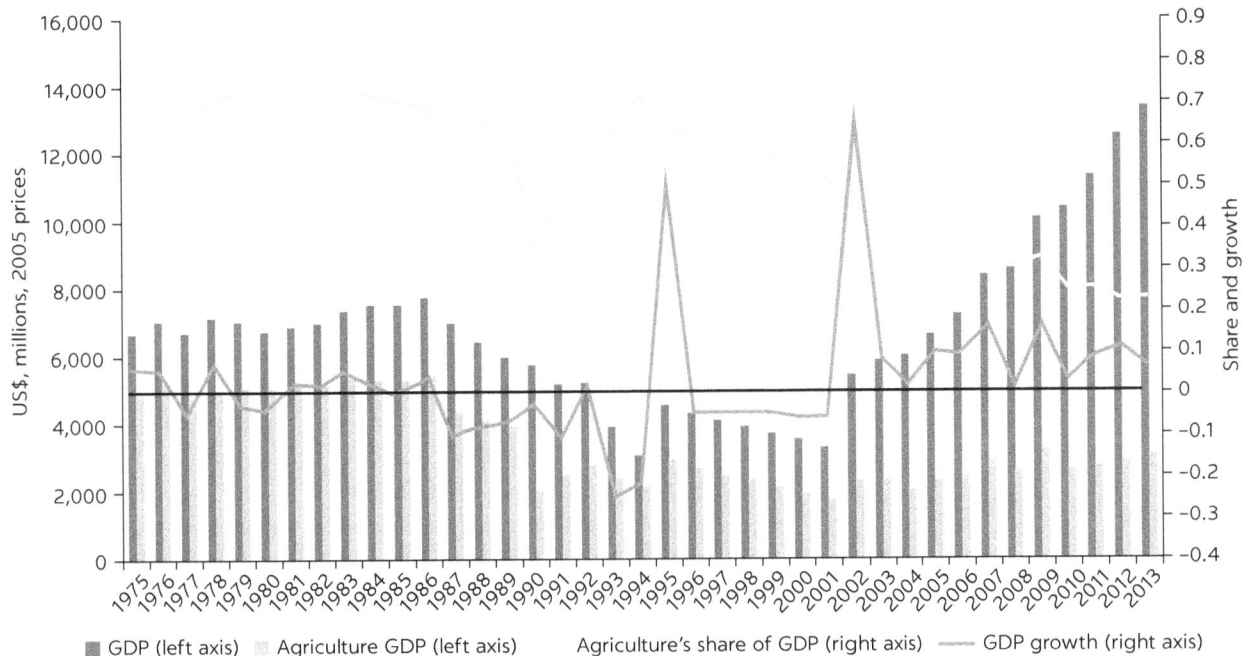

■ GDP (left axis)　░ Agriculture GDP (left axis)　━ Agriculture's share of GDP (right axis)　┈ GDP growth (right axis)

Source: World Bank 2016c.

decline in the sector's share of total GDP, from 71 percent in 1994 to 24 percent in 2013 (figure 1.3). Despite this, the sector employs 40 percent of the total labor force, and more than half of the rural workforce is involved in agriculture. Moreover, an analysis based on household survey data from the NRVA 2011–12 and ALCS 2013–14 demonstrates an increase in agriculture's employment share in rural areas. About 70 percent of the population lives and works in rural areas, mostly on farms, and 61 percent of all households derive income from agriculture. Of off-farm employment, including in urban and peri-urban areas, a large share of employment is in agriculture-related sectors and food processing, and agricultural industry accounts for most of the exports and about 41 percent of manufacturing.

As development proceeds in Afghanistan, the number of people employed directly in primary agricultural production will decline and the sector will naturally shed labor to other sectors, particularly services and manufacturing, but also to agribusiness and the agro-food system. For the near future, however, and especially as the economy adjusts to changes brought about by the decline of aid flows and foreign military expenditures, the agriculture sector's ability to absorb new farm-level workers while raising farm productivity will be crucial for poverty reduction and economic growth. With nearly 70 percent of Afghans perceiving poverty and unemployment as the major cause of conflict (Oxfam 2009), the need for leveraging the agriculture sector for job creation, sustainable development, food and nutrition security, and women's economic empowerment is crucial to long-term security. As opportunities are developed to increase the commercialization of on-farm and agribusiness products and enhance productivity, agriculture has the potential to grow into a major sector to support economic growth and more productive, sustainable jobs for both men and women. In a well-developed and mechanized agriculture sector, jobs move up in the value chain, rather than out.

RELEVANCE OF AGRICULTURE TO JOB CREATION AND LABOR MARKETS

Agriculture is the world's largest provider of jobs. Multidimensional development policies are needed to create new and inclusive jobs in the sector and improve productivity of existing practices. In Afghanistan, a mostly dry country, wheat dominates agriculture in the irrigated plains. In dry zones, people are mostly involved in fruit and livestock. Therefore, technical and financial support for expansion of irrigation facilities and adoption of agricultural technologies may benefit wheat producers in the irrigated plains, and support for horticulture and livestock may benefit farmers in the dry zones. Improvements in finance and market access, however, may benefit all farmers.

High growth in agriculture and the agro-food system can help raise labor demand in the rural labor market, reducing unemployment and underemployment. Agriculture and the agro-food system can also support most factors that have a positive impact on labor demand in rural markets (figure 1.4).

Most supply-side factors for rural labor markets increase total labor supply, but it is crucial to catalyze agricultural productivity and agricultural value chains. An increase in rural labor supply due to the female labor force participation rate, literacy, and internally displaced people and returnees pressures labor markets, and an excess supply leads many to be unemployed or underemployed. High unemployment in rural areas, particularly among

FIGURE 1.4

Framework for job creation in rural Afghanistan

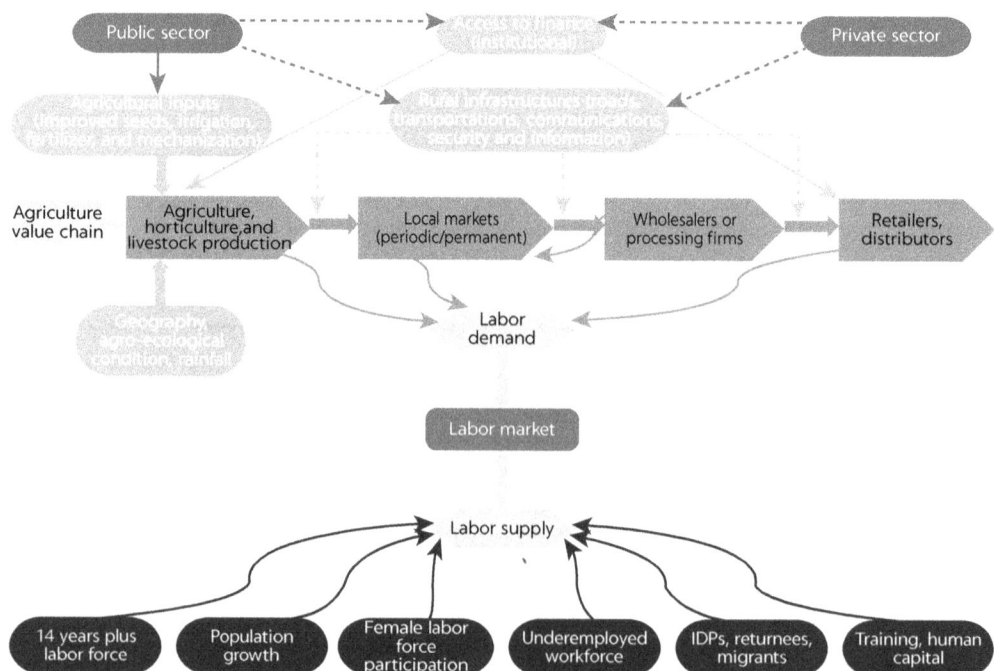

Note: IDP = internally displaced persons.

women and youth, cannot be tackled if there are no increases in the demand for labor through improved productivity of crop agriculture, diversification toward high-value agriculture (e.g., horticulture and livestock), or linkages between farmers and the markets and the rural nonfarm economy. Both the public and private sectors can support rural jobs through effective policy design and investments in agricultural inputs, rural infrastructure, and access to finance.

APPROACH AND METHODOLOGY

This report assesses employment support and creation in Afghanistan through agriculture. It uses a multidimensional approach to analyze jobs in the sector. First, it evaluates the current employment structure in rural areas to better understand the kind of work people pursue and the challenges they face in securing their livelihoods. Second, it analyzes the nature of employment, skills, and human capital of vulnerable groups such as women, youth, and the landless, covering the bottom 40 percent of income earners to understand the inclusiveness of rural jobs. Third, it evaluates the role of public and private sector interventions in supporting job creation and linking farmers to markets.

The study used an empirical method to keep the analysis as objective and factual as possible. The analysis therefore relied on many data sources, including a new survey of market participants along the value chains for selected high-value products. We used national- and province-level aggregate data from Afghanistan's CSO for national accounts statistics and agricultural production, along with agricultural data from the Food and Agriculture Organization of the United Nations. Household-level data sources included nationally and provincially representative surveys from the NRVA 2011–12 and ALCS 2013–14. We used beneficiary data and other information from project offices to measure employment impacts. The micro-level datasets provided most of the data used to obtain the original empirical findings; for details on the datasets, see annex 1A.

The report also measures job creation from the World Bank's agriculture projects in Afghanistan. The complexity of the job creation concept makes jobs measurement challenging and prone to errors (Fowler and Markel 2014), especially in the agriculture sector. The primary concerns of measuring an agriculture project's job creation impact are formal and/or informal jobs; jobs for rural target groups; sustainable and/or temporary/ seasonal jobs; jobs for the underemployed and unpaid family workers; and jobs for those who are already occupied with low-productivity tasks. It is also important to note that agriculture projects can create direct, indirect, and induced jobs. There are many ways to measure jobs creation in agriculture (as could be seen in annex 1A) and this report adopts the FTE approach. Under this approach, net additional jobs created by a program are transformed into FTE jobs by dividing the net extra labor days of work generated from implementation by a specific agriculture project. While the number of labor days for a FTE job varies in jobs measurement literature, this report considers 200 labor days of work as a FTE job. Therefore, the total number of FTE jobs created can be measured by dividing the net additional days of work due to program intervention by 200. While this helps reveal the depth of job creation by estimating the number of labor days created, it does not measure the number of people who employed by the program.

AFGHANISTAN ECONOMIC PERFORMANCE IN A REGIONAL CONTEXT

Every two out of five rural people in Afghanistan live under the poverty level, with insufficient earnings to satisfy basic food consumption and nonfood needs. Between 2008 and 2012, although the urban poverty rate declined about 1 percent, the rural poverty rate increased slightly (World Bank 2015b). Poverty rates remained stagnant despite overall strong economic growth and a substantial increase in international aid in security and reconstruction. Afghanistan's poverty incidence is among the highest in Asian developing countries, and its headcount poverty—the percent of population below the national poverty line—in rural and urban areas is among the highest of its neighboring countries. In terms of rural poverty, however, it is not far behind Bangladesh, Pakistan, or Tajikistan. Yet, its urban poverty is much higher than all neighboring countries, except Tajikistan (figure 1.5).

From 2008–12, Afghanistan's per capita constant-price GDP (the value of goods and services in relation to a base year) grew at a rate similar to neighboring countries (figure 1A.1). However, poverty was stagnant in the same period due to an increase in inequality; the Gini index, which represents the income or wealth distribution of a nation's residents, increased from 29.7 in 2008 to 31.6 in 2012 (World Bank 2015b). Since 2012, all neighboring countries experienced continuous growth in per capita GDP, but it declined slightly in Afghanistan, as did per capita real GDP, from $651 in 2012 to $624 in 2015. In the same period, per capita GDP in Bangladesh, Pakistan, and Tajikistan increased by $121, $96, and $88, respectively.

Afghanistan's GDP composition has changed significantly since the fall of the Taliban. International aid-based reconstruction efforts have helped the service sector to grow faster than agriculture and industry, its share of total GDP increasing from 39 percent in 2003 to 55 percent in 2013. Agriculture's share of total GDP declined from 38 percent in 2003 to 24 percent in 2013 (figure 1.6). The share of manufacturing remained relatively stable, though the sector contracted in the past decade.

While Afghanistan's labor force participation rate is the lowest in South Asia, its unemployment rate is the highest (figure 1.7). (Its unemployment rate is lower

FIGURE 1.5

Poverty headcount ratios in Afghanistan vis-à-vis neighboring countries

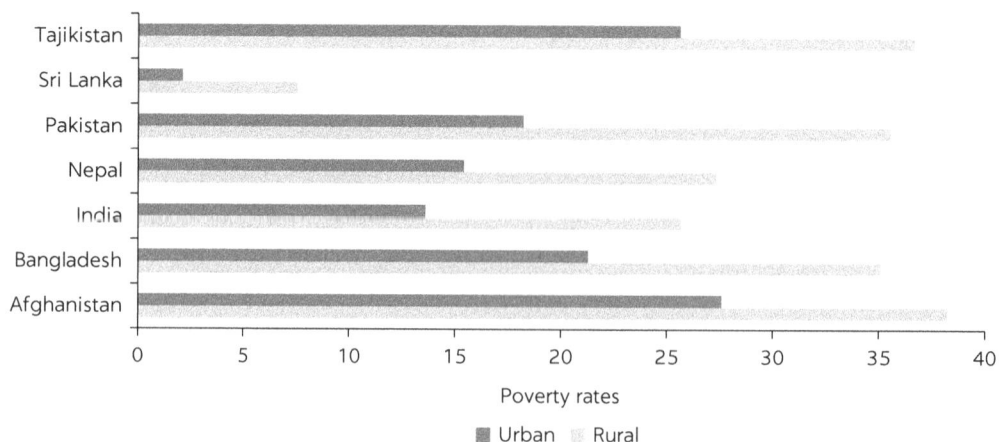

Poverty rates

■ Urban   Rural

Source: World Development Indicators 2016. Statistics are from 2010 or later.

FIGURE 1.6

Composition of gross domestic product: Afghanistan vis-à-vis neighboring countries

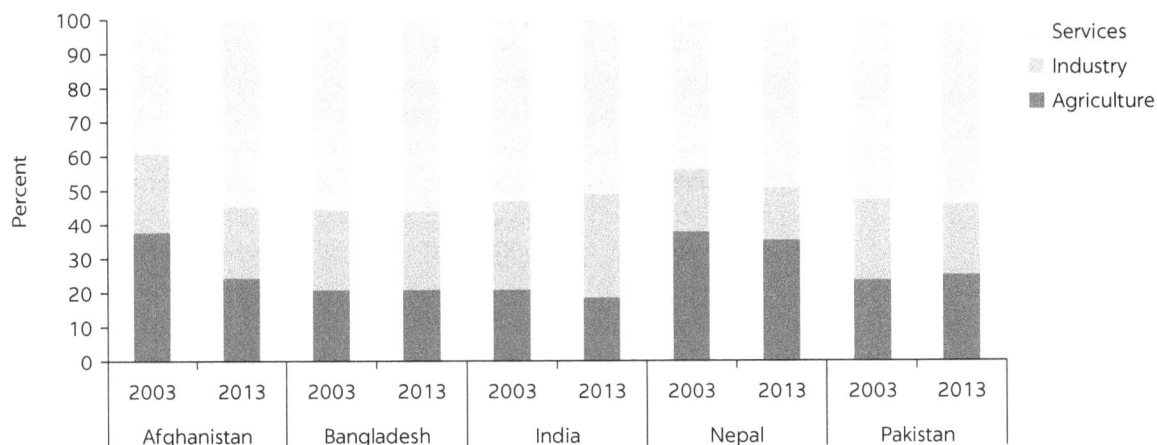

Source: World Bank 2016c.

FIGURE 1.7

Unemployment and labor force participation rates in Afghanistan versus neighboring countries (average 2006–15)

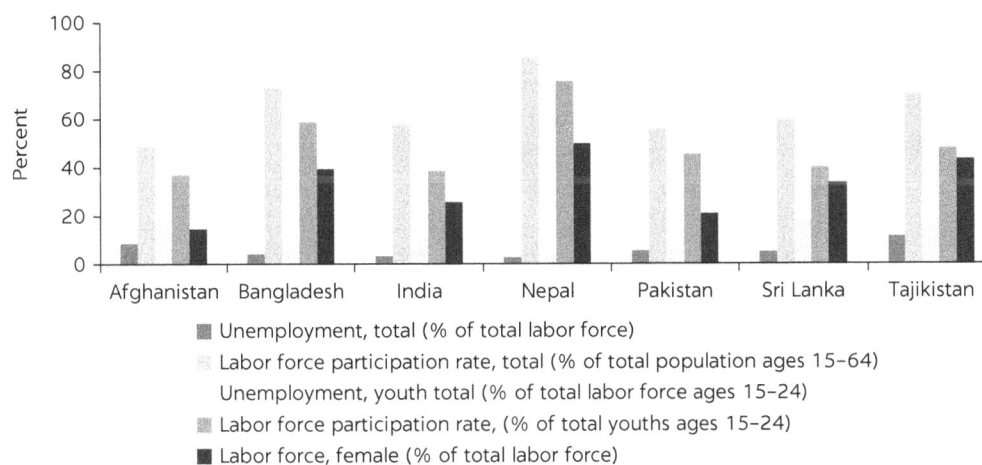

Source: World Bank 2016c.

than Tajikistan's, however.) Youth unemployment is the highest among neighboring countries, though its youth labor force participation rate is close to India's and Sri Lanka's. One of every five Afghan workers aged 15–24 who participated in the labor force remained unemployed. The youth unemployment rate in is about 2.5 times higher than Pakistan's. Many working-age women are out of the labor force, as female workers account for just 15 percent of total labor force. This is the lowest rate in the region, though it is not much different in Pakistan.

Afghanistan is also far behind its neighbors in literacy. While most neighboring countries achieved youth literacy rates above 80 percent (Pakistan reached more than 70 percent), Afghanistan's is about 47 percent (figure 1A.2). The literacy rate among adult males is not encouraging in the regional context, and the adult female literacy rate is even more severe. To illustrate, it is less than half the rate in Pakistan.

ANNEX 1A: MICRODATA USED IN THE ANALYSIS—NRVA 2011–12 AND ALCS 2013–14

This report made extensive use of the NRVA 2011–12 and ALCS 2013–14. These surveys were part of the NRVAs conducted across Afghanistan since 2003. (The earlier surveys were known as "National Risk and Vulnerability Assessments," but the 2013–14 iteration was called the "Afghanistan Living Condition Survey" to reflect the complete scope of the development information it covered.) Their methodologies considered the country circumstances, and were designed to comply with international survey recommendations and best practices. The ALCS has maintained a degree of consistency in design, sampling, and questionnaires to ensure comparability over time. However, application of the international standards resulted in some methodological changes in the ALCS 2013–14, which restricts the comparability of results with the results from the NRVAs.

While both surveys covered areas such as household demographics, agricultural and nonagricultural economic activities, labor market behaviors, and migration situations, some sections differed. The most important changes were made in the labor market module. In the ALCS 2013–14, labor market indicators were collected through a detailed module, whereas the NRVA 2011–12 used an abbreviated module. Due to these changes, it was difficult to compare their labor market outcomes. Accordingly, we emphasized the spatial and structural analysis of labor market outcomes rather than the dynamic trend analysis of labor market indicators.

Both surveys followed a similar sampling design and ensured survey representativeness at the national and provincial levels, as well as for *Hijri Shamsi* (solar Islamic) calendar seasons, Afghanistan's official calendar year, the first day of which is March 21. Both surveys identified 35 strata, 34 for the provinces and 1 for the nomadic Kuchi population. Data collection in all provinces was distributed equally over the 12 months to achieve stratification by season, a standard procedure to attain representative statistics in different seasons. The NRVA 2011–12 used the pre-census household listing conducted by the CSO in 2003–05; the ALCS 2013–14 used the one conducted in 2003–05 and updated in 2009. Both surveys covered 20,400 sample households across Afghanistan.

Concepts and definitions for job creation in agriculture

There is consensus among policymakers, development practitioners, and economists about the role of productive employment in poverty reduction. Job creation has been central to development discourse, and international development communities are increasingly orienting their investments toward job creation. While some programs are aimed at providing safety nets to the poor, others are more integrated into countries' development programs and designed to generate new employment opportunities for target groups. Projects that promote job creation are an important policy instrument, especially in fragile, post-conflict economies where unemployment and underemployment rates are high, and security concerns undermine livelihoods and require income-stabilizing interventions.

However, the complexity of the job creation concept makes measurement challenging and prone to errors (Fowler and Markel 2014), especially in the agriculture sector. Thus, it is necessary to understand the key terms related to job creation to estimate the impacts of development projects. The primary concerns of measuring an agriculture project's job creation impact are formal and/ or informal jobs; jobs for rural target groups; sustainable and/or temporary/seasonal jobs; jobs for the underemployed and unpaid family workers; and jobs for those who are already occupied with low-productivity tasks. It is also important to note that agriculture projects can create direct, indirect, and induced jobs. The literature notes different ways of measuring job creation, such as these from Fowler and Markel (2014):

Full-time equivalent: The number of FTE jobs as a measure of job creation is often estimated in employment impact assessments. Under this approach, net additional jobs created by a program are transformed into FTE jobs by dividing the net extra labor days of work generated from implementation by a specific number of labor days. For example, the Donor Committee on Enterprise Development standard of measuring job creation considers a full-time job as 240 days of work in a year. Therefore, the total number of FTE jobs created can be measured by dividing the net additional days of work due to program intervention by 240. Using this method, if a program generates a job that requires 120 days of work, then the project, in practice, generates 0.5 FTE jobs. While this helps reveal the depth of job creation by estimating the number of labor days created, it does not measure the number of people who actually benefitted from the program.

Direct measurement and employment multiplier: Direct measurement may be done by analyzing employer records, conducting employer surveys, or surveying employees. Employment multipliers may be applied by developing localized multipliers or estimating the employment elasticity of project-generated income. To estimate employment multipliers, it is necessary to collect information from market actors to develop a localized employment multiplier, then calculate the direct and indirect job creation. Another approach is to use published employment elasticity figures to estimate a program's impact on employment. This is useful for measuring induced jobs when there are published, credible employment multipliers. Input–output tables can be used to calculate the induced effects on other sectors by linking project-level outputs to sectors that link sector wage bills to sector employments, generating the relation between the changes in the value of a sector's output and changes in the number of employed. A social accounting matrix may be appropriate to measure employment generation, while input–output tables can be helpful in countries where a matrix does not exist.

International Labour Organization method: With this method a person is employed if he or she works during a specified period (e.g., one week). While this helps to identify the number of program beneficiaries, it fails to inform the depth and sustainability of jobs created. It can also obscure the total quantity of work created, because it measures the work *period*, not work *hours*. Furthermore, it does not capture the quality of the job that is created.

U.K. Department for International Development (DFID) job headcount approach: This approach (DFID 2012) includes a job headcount indicator, which considers a job to be a person who works at least 20 hours per week for at least 26 weeks in a year; works in conditions that comply with the International Labour Organization's eight Core Conventions; and earns the greater of the country's national minimum wage or the wage required to take the household's members to the $1.25 poverty line. Because it combines three important estimation issues—the number of people benefitting from job improvement, the quality of the job in terms of wages and rights, and the increase in income for existing workers—it is helpful in contexts where significant numbers of people are working full time but gaining little from their labor. Regardless, by setting a time threshold for counting a job, it cannot demonstrate the depth of employment generation.

It is also important to identify who is benefitting from job creation (e.g., targeted poor communities or farmers). Depending on the type of intervention, nontargeted groups may be benefitting by indirect or induced effects. Often, projects consider measuring and reporting only direct job creation, yet their impacts on indirect or induced job creation may be much greater, as is the case of agricultural irrigation projects. Fowler and Markel (2014) define these three types of job creation as follows:

1. **Direct job creation** is net additional employment created in a sector as a result of program implementation. It counts the jobs created by service providers or producers working directly with the program. It also considers the creation of temporary jobs. Direct employment may be temporary or sustainable.

2. **Indirect job creation** is the additional employment generated (or lost) as other sectors respond to the intervention by expanding their output to supply inputs and outputs.

3. **Induced job creation** stems from interventions that may increase farm households' income, which in turn raises demand for outputs and services from other sectors. The other sector also increases its employment. Induced effects can be calculated by linking extra household income generated to household spending in each sector, and the direct and indirect employment required for this additional production. Both direct and indirect job creation would raise expenditures on consumption, education, and health, which would increase further employment.

Direct, indirect, and induced employment can fall into six categories:

1. **Temporary/seasonal employment:** This occurs during a limited or defined period in a year. It is characteristic of agricultural employment, as additional demand for labor arises during certain stages of production (e.g., land preparation and harvesting), not continuously throughout the year. Programs and policies that boost agriculture can create temporary/seasonal employment for the rural poor, and programs should include these positions in job creation measurements.

2. **Formal employment:** Work in the private or public sector as salaried workers is considered formal employment. It often requires more education and better skills. It is a more sustainable form of employment generation and provides more benefits and job security.

3. **Informal employment:** This includes work in agriculture, wage labor, and nonfarm small businesses. It requires less skill, and employees are more vulnerable because they are not protected by regulation. Short-term job creation programs and policies usually focus on jobs in the informal economy.

4. **Wage employment:** This includes adult workers in households who work for daily wages.

5. **Self-employment:** The NRVA and ALCS categorize the self-employed as adult household members who say they work as farmers, sharecroppers, shop owners, street vendors, independent workers, and the like.

6. **Unpaid family work:** Unpaid family workers are adult household members who voluntarily work in their family farms or businesses.

FIGURE 1A.1

Gross domestic product per capita: Afghanistan vis-à-vis its neighbors

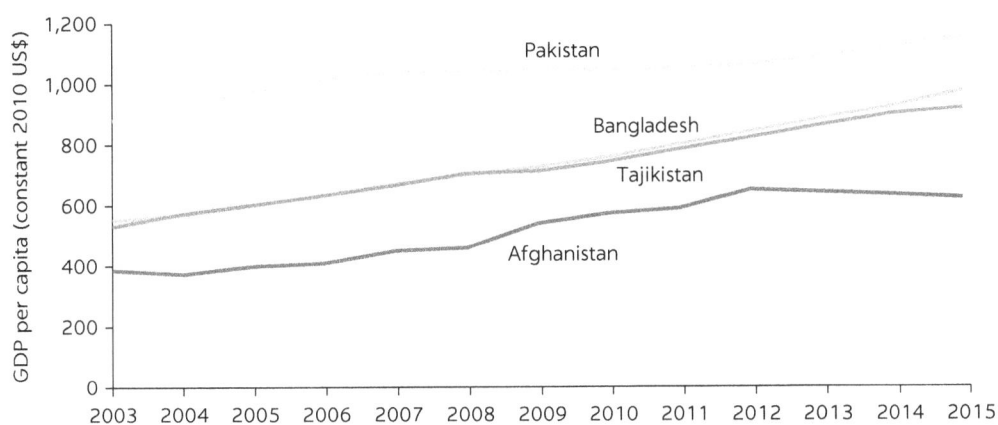

Source: World Development Indicators 2016.

FIGURE 1A.2

Regional literacy rates in Afghanistan vis-à-vis neighboring countries, average, 2006–15

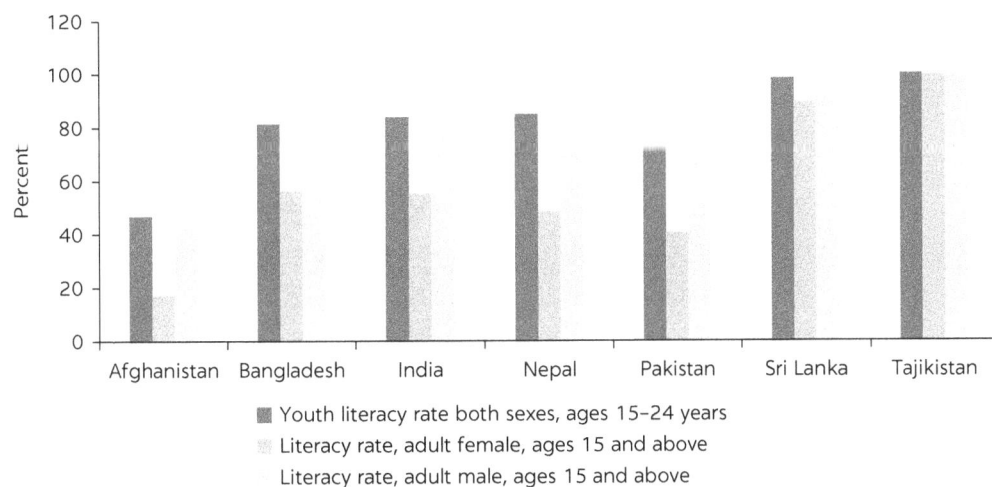

■ Youth literacy rate both sexes, ages 15–24 years
▨ Literacy rate, adult female, ages 15 and above
  Literacy rate, adult male, ages 15 and above

Source: World Development Indicators 2016.

Afghanistan's official definition of employment indicators (Redaelli 2013), includes the following:

- **Employed:** working for more than 8 hours
- **Unemployed:** working fewer than 8 hours per week, not working, looking and available
- **Underemployed:** working more than 8 hours but fewer than 40 hours per week, willing to work more hours and available
- **Not gainfully employed:** unemployed or underemployed

NOTES

1. The ASR proposes "a 'first-mover' strategy to serve as an initial phase in the national agricultural strategy, providing an early boost to productivity growth, employment, and poverty reduction. The promotion of a 'first mover' strategy responds both to the adjustment of the anticipated decline in foreign aid and agricultural transformation that is necessary for Afghanistan's inclusive economic growth, job creation, and food security" (World Bank 2014a, x–xi).
2. Opium remains a significant economic and political factor. Roughly one-third of Afghanistan's GDP is estimated to come from opium. The U.S. government has engaged in significant eradication efforts, yet the returns to opium production outpace those of most other agricultural activities. Land under opium continues to grow; the total area under opium poppy cultivation was estimated at 224,000 hectares in 2014, a 7 percent increase from 2013. Potential opium production was estimated at 6,400 tons in 2014, an increase of 17 percent from its 2013 level (5,500 tons) (UNODC 2014).

REFERENCES

CSO (Central Statistics Organization). 2016. "Demographic and Social Statistics." CSO, Kabul.

Davis, S., and J. Haltiwanger. 1992. "Gross Job Creation, Gross Job Destruction, and Employment Reallocation." *The Quarterly Journal of Economics* 107 (3): 819–63.

DFID (Department for International Development). 2012. "How to Note: Measuring Job Creation 2: How Do We Define a Job?" DFID, UK.

Fowler, B., and E. Markel. 2014. "Measuring Job Creation in Private Sector Development." Working Paper, Market Share Associates, Ottawa, Ontario, Canada.

OECD (Organization for Economic Co-operation and Economic Development). 2014. "Employment Policies and Data." OECD, Paris, France.

Oxfam. 2009. *The Cost of War: Afghan Experiences of Conflict, 1978–2009.* Oxford, UK: Oxfam International.

Redaelli, S. 2013. "Developing a Job Creation Monitoring Framework for the Afghanistan Portfolio." Presentation, December.

UNODC (United Nations Office on Drugs and Crime). 2014. "Afghanistan Opium Survey 2014." UNODC, Vienna, Austria.

World Bank. 2005. "Afghanistan: National Reconstruction and Poverty Reduction—The Role of Women in Afghanistan's Future." World Bank, Washington, DC.

———. 2011. "Understanding Gender in Agricultural Value Chains: The Cases of Grapes/Raisins, Almonds and Saffron in Afghanistan." World Bank, Washington, DC.

———. 2012. *World Development Report 2013: Jobs.* Washington, DC: World Bank.

———. 2013. "Afghanistan Economic Update." World Bank, Washington, DC.

———. 2014a. "Afghanistan Agriculture Sector Review (ASR)." World Bank, Washington, DC.

——. 2014b. "Women's Role in Afghanistan's Future: Taking Stock of Achievements and Continued Challenges." World Bank, Washington, DC.

——. 2015a. "Afghanistan Country Snapshot." World Bank, Washington, DC.

——. 2015b. "Afghanistan Poverty Status Update." World Bank, Washington, DC.

——. 2016a. "Afghanistan Systematic Country Diagnostic (SCD)." World Bank, Washington, DC.

——. 2016b. "Navigating Risk and Uncertainty in Afghanistan." Technical Brief. World Bank, Washington, DC.

——. 2016c. "World Development Indicators 2016." World Bank, Washington, DC.

Yousufi, A. 2016. "Horticulture in Afghanistan: Challenges and Opportunities." *Journal of Developments in Sustainable Agriculture* 11 (1): 36–42.

.

2 Employment Patterns in Rural Afghanistan

INTRODUCTION

This chapter discusses the current employment structure and spatial patterns of rural employment in Afghanistan to better understand the nature of the work people pursue and the challenges they face in securing their livelihoods. It pays particular attention to inclusive aspects of rural jobs. It uses employment data from the Afghanistan Living Condition Survey (ALCS) 2013–14, as well as data from the National Risk and Vulnerability Assessment (NRVA) 2011–12 and secondary data.

The chapter also explores the patterns of earnings from rural activities. Employment patterns across sectors provide insights into the roles each plays in overall rural employment. Because the quality and sustainability of the jobs in one sector often cannot be assessed properly by relying only on the employment patterns across sectors, we have also analyzed income data to explore the sectors that strongly support income generation. Comparing employment and income patterns in different sectors allows us to better understand returns from jobs, which enables policymakers to better formulate and implement policies.

THE NATURE, TYPES, AND SECTORAL DISTRIBUTION OF RURAL EMPLOYMENT

This section explores the changing age structure of Afghanistan's rural workforce, the spatial pattern of the labor force participation rate (LFPR), rural unemployment and underemployment, and the distribution of rural workers between the broad sectors of agriculture and non-agriculture.

The rural workforce's age structure

The age structure of Afghanistan's rural population (figure 2.1 and figure 2A.1) reveals an impending "youth bulge" in the labor force and indicates the challenges for designing and formulating sound and effective job creation policies to cope with it. About 46 percent of the rural population is younger than 14, and more people will be entering the workforce in the coming years.

FIGURE 2.1

Population age structure in rural Afghanistan

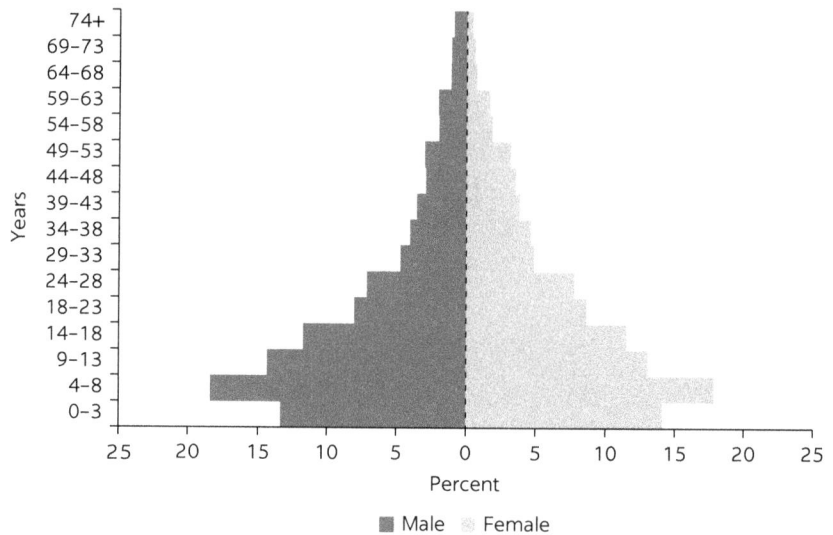

Source: Based on ALCS 2013–14.
Note: ALCS 2013–14 data show the age-structure for those years. For 2018–19, it is assumed that the LFPR for males and females will remain constant and that children ages 9–13 in 2013–14 will enter the workforce in 2018–19.

FIGURE 2.2

Employment structure in rural Afghanistan

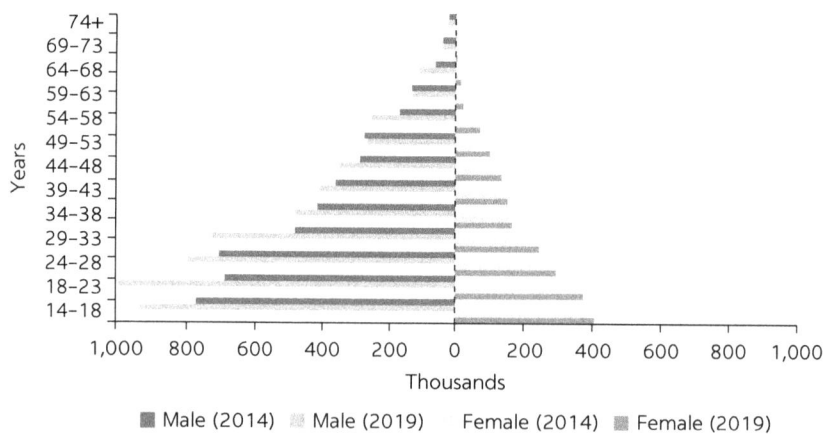

Source: Based on ALCS 2013–14.

Figure 2.2 shows that more and younger people will be of working age within the next five years. There were 1.4 million male rural workers younger than 23 in 2013–14; that will increase to 1.9 million in 2018–19, adding about 500,000 young male workers to the workforce. About 160,000 more female workers will be in the workforce in the next five years. Therefore, not only do current deficits related to jobs need to be addressed, more than 600,000 extra new jobs will need to be created in the next two years to ensure employment and livelihoods for the growing young workforce in rural areas.

The number of young people entering the labor force each year is much higher than the vacant positions left by their older cohorts, implying an

FIGURE 2.3

Schooling of workforce in rural Afghanistan, 2013–14

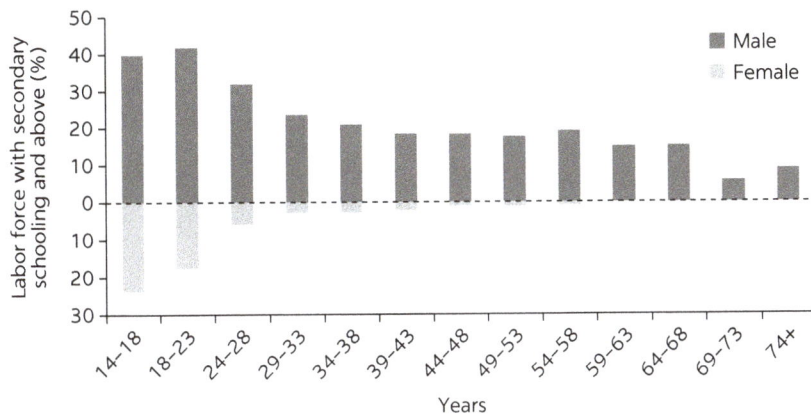

Source: Based on ALCS 2013–14.

excess supply in the rural labor market, which will lead to competition for each new job. Moreover, the new generation is equipped with higher human capital (figure 2.3) and is likely to be more literate than their elders.

For example, more young male workers aged 18–23 have some secondary or higher schooling than male workers aged 23–28 (41.8 percent versus 31.9 percent). Young females in the workforce also have more education than their elder cohorts. Thus, the educated youth workforce will be aiming for higher-skilled jobs than past generations—which means policymakers and development practitioners must generate new, appropriate jobs. Failure to create better and more inclusive jobs may frustrate young workers, resulting in further social instability.

Labor force participation in rural Afghanistan

Generally, about half of the working-age population in rural areas participates in the labor force. Female participation is generally low. Yet, the LFPR is dispersed across rural provinces (map 2.1), with a comparatively high LFPR in Wardak and Paktika. The map's unemployment panel shows high rates in Ghor and Daykundi provinces; its underemployment panel reveals that employed people who work less than 40 hours a week can account for more than 60 percent of total employed people in some provinces (for example, Helmand, Wardak, and Zabul). Map 2.1 also illustrates underemployment and unemployment rates across rural regions.

As figure 2.4 illustrates, the average LFPR was 55 percent in 2013–14, but it varied substantially across regions. It was above 60 percent in the north, south, and west central regions, but was much lower in the northeast (43.5 percent) and the southwest (46.8 percent).

The unemployment situation is alarming across rural Afghanistan. One in every five adults who participates in the labor force and is willing to work is unemployed (figure 2.5). The unemployment rate is severe in the west central region (about 42 percent) and west region (about 31 percent). However, in the central region, which includes Kabul, it is also above the

MAP 2.1

Labor force participation, employment, unemployment, and underemployment in Afghanistan, 2013–14

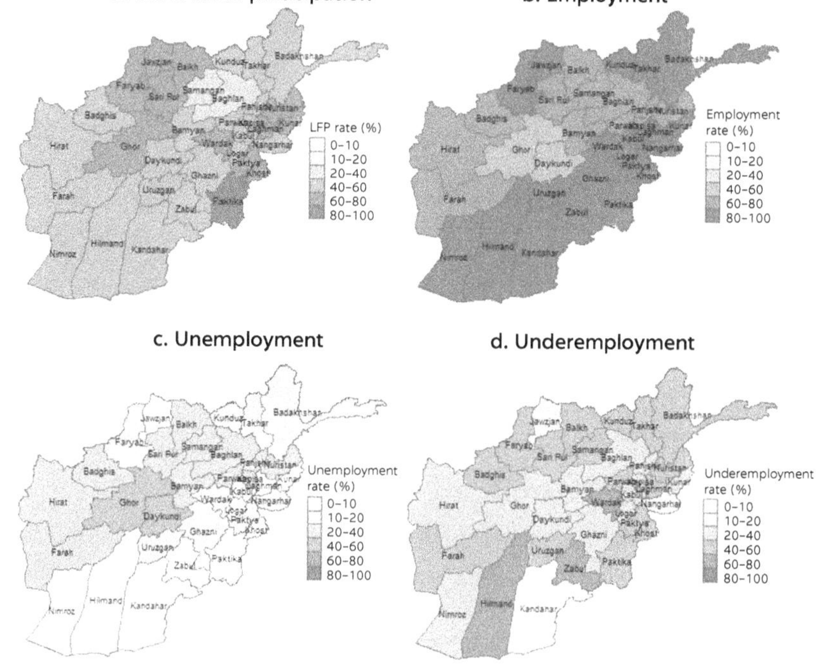

Source: Based on ALCS 2013–14.

FIGURE 2.4

Labor force participation rates in rural Afghanistan, 2013–14

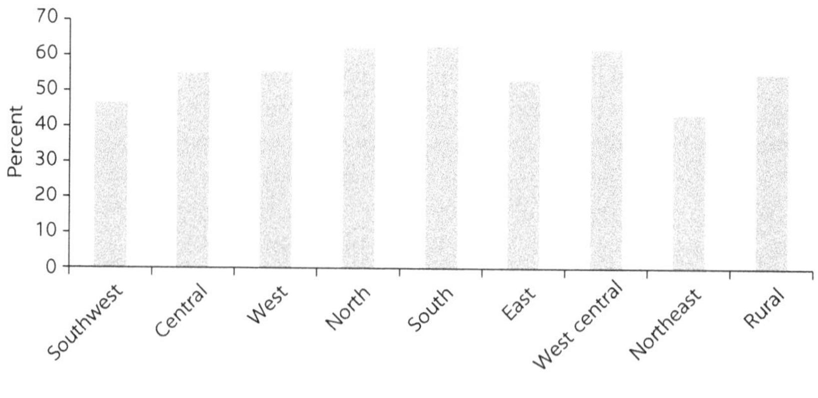

Source: Based on ALCS 2013–14.

rural average, implying that rural people living near urban areas also lack sufficient employment opportunities.

Figure 2.5 also shows that underemployment in rural areas, at about 52 percent, is a serious concern. The underemployed are those who work more than 8 hours but fewer than 40 hours a week. Underemployment varies significantly across regions. It is highest in the west (63.4 percent) and the northeast

FIGURE 2.5

Underemployment and unemployment rates in rural Afghanistan, 2013–14

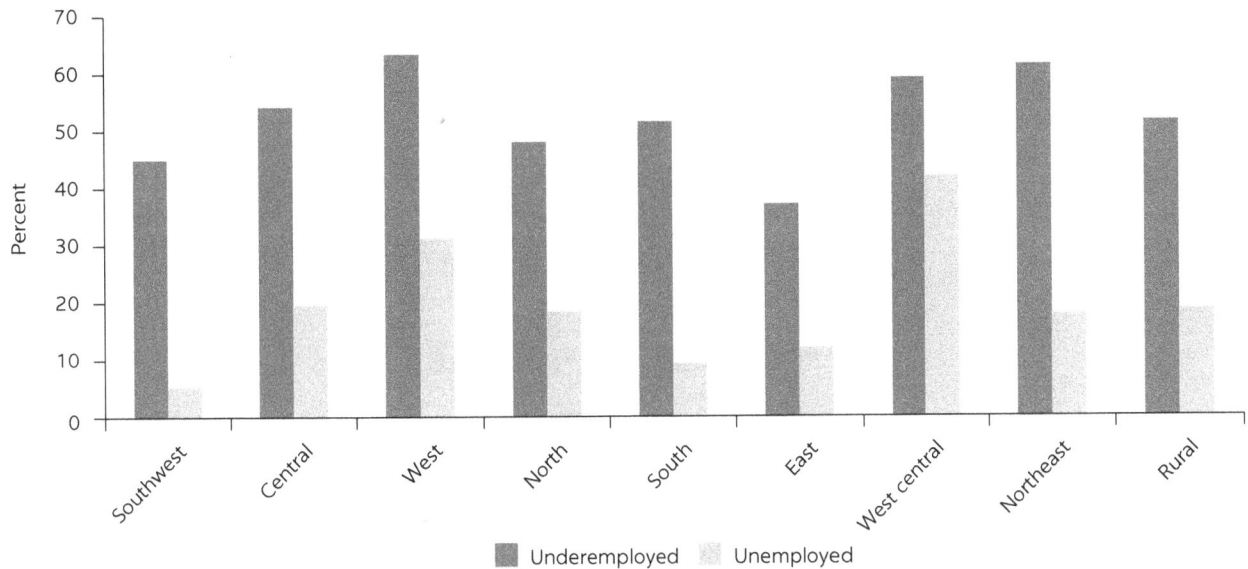

Source: Based on ALCS 2013–14.

FIGURE 2.6

Annual per capita income and income growth in rural Afghanistan, 2011–12 and 2013–14

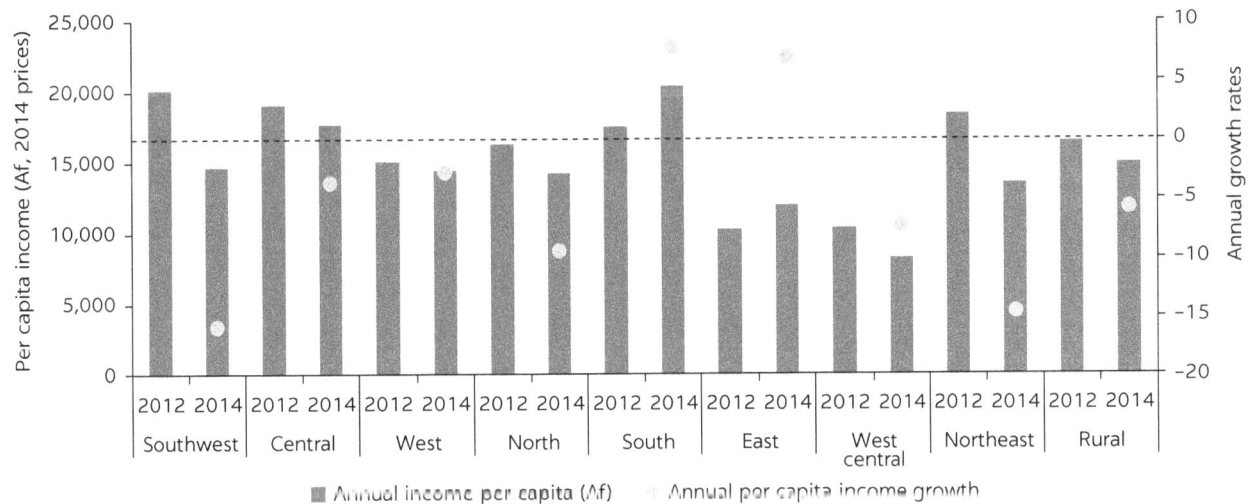

Source: Based on NRVA 2011–12 and ALCS 2013–14.

(61.6 percent), and is severe in the west central region (59.2 percent), which also has the highest unemployment. There, two of five people in rural areas are unemployed, and those who do find jobs are likely to be underemployed. Underemployment is also high in the central region.

Income data provides some insights about rural income dynamics in recent years.[1] Figure 2.6 presents average annual per capita income of rural people for 2011–12 and 2013–14, showing an average decline of about 5 percent and a steeper decline (about 15 percent) in the southwest and northeast. The high decline in the northeast aligns with the region's high unemployment.

The east and the south, which apparently have lower unemployment rates than most other regions, enjoyed positive annual income growth of about 7 percent. The southwest had the worst decline, yet had the lowest unemployment rate. Regional differences can provide more insights about this paradox: Because rural people in the southwest are primarily involved in agriculture, they are less likely to be unemployed; however, workers in the sector experienced a decline in income due to a falling crop prices in recent years. Figure 2A.3 highlights the substantial variation in per capita annual income in 2013–14 across provinces. Per capita annual income is about Af 26,735 in Kabul province; it is 75 percent less (Af 6,610) in Uruzgan.

Rural employment: farm versus nonfarm

Agriculture's role in total gross domestic product (GDP) is declining. Its share of total GDP was 71 percent in 1994, but was less than 25 percent in 2013. Still, more than half of the rural workforce is involved in agriculture, which means labor productivity in the sector has been falling behind other sectors. Higher labor productivity in nonfarm sectors is expected to motivate agriculture workers to pursue better-paying nonagricultural jobs in rural areas. Evidence suggests that employment in the agriculture sector has increased in recent years (figure 2.15). For example, 4.5 million people worked in agriculture in 2000–05, but that had risen to 5.6 million in 2011–13. Figure 2.7 also reveals that, while the agricultural employment share in rural areas is about 54 percent, agricultural income share is only about 36 percent. This underscores that rural nonfarm workers earn more than rural farm workers. Relatedly, many rural Afghans are involved in nonagricultural activities, mainly as day laborers.

The low share of agricultural income can be attributed to three major facts: a) non-imputation of subsistence consumption in income, b) low market participation, and c) high number of unpaid family workers. While the

FIGURE 2.7

Rural employment and income share: agriculture vs. non-agriculture, 2013–14

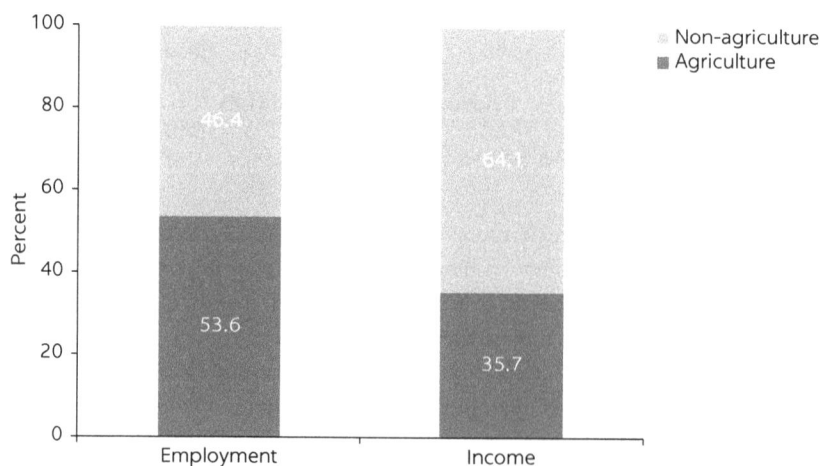

Source: Based on ALCS 2013–14.

unpaid family workers account for less than 10 percent of nonfarm employment, they are about 54 percent of agricultural employment (figure 2.8), or about 1.4 million workers. This suggests that mechanisms are needed to transition this group to paid work. One explanation for the high numbers of unpaid family workers in rural agriculture is the influx of youth into the labor force. Without opportunities in the nonfarm sector, these young workers are engaging as unpaid family workers.

Map 2.2 presents the patterns of agriculture's employment and income share in rural areas in 2013–14. Its income share is generally high in the southern provinces, while it is generally less than 40 percent in the northern provinces. It is even less in Kabul and its surrounding provinces, consistent with what is to be expected in urban areas.

Figures 2.9 and 2.10 plot the spatial pattern of employment and income share from agriculture and non-agriculture at the regional level. Figure 2.10 reveals that shares from agriculture vary from 25 percent to 69 percent. In the central, south, and east regions around Kabul, the income share ranges from 20 percent to 30 percent. Agriculture's employment share is also less in these regions, consistent with what is to be expected in urban or peri-urban areas. The large share

FIGURE 2.8

Agriculture vs. non-agriculture employment patterns, 2013–14

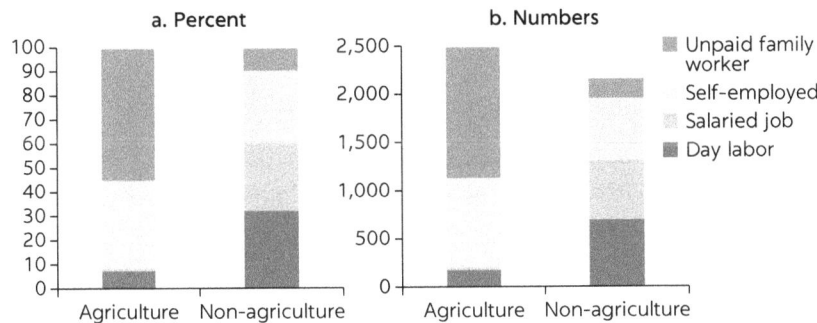

Source: Based on ALCS 2013–14.

MAP 2.2

Spatial pattern of income and employment in agriculture, 2013–14

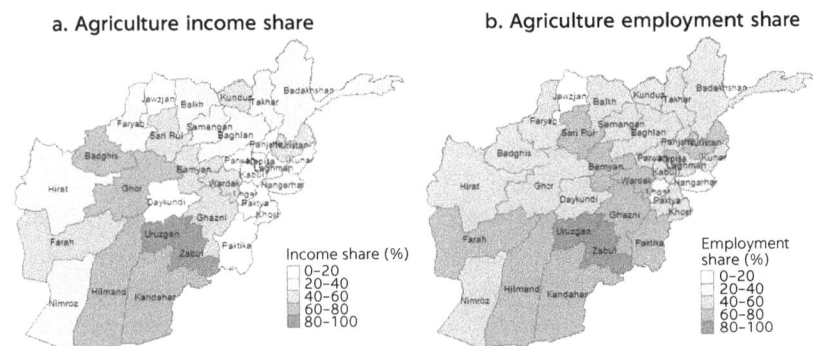

Source: Based on ALCS 2013–14.

FIGURE 2.9
Income shares and spatial pattern of income: regional level

a. Income share

b. Income (Af) in 2014 prices

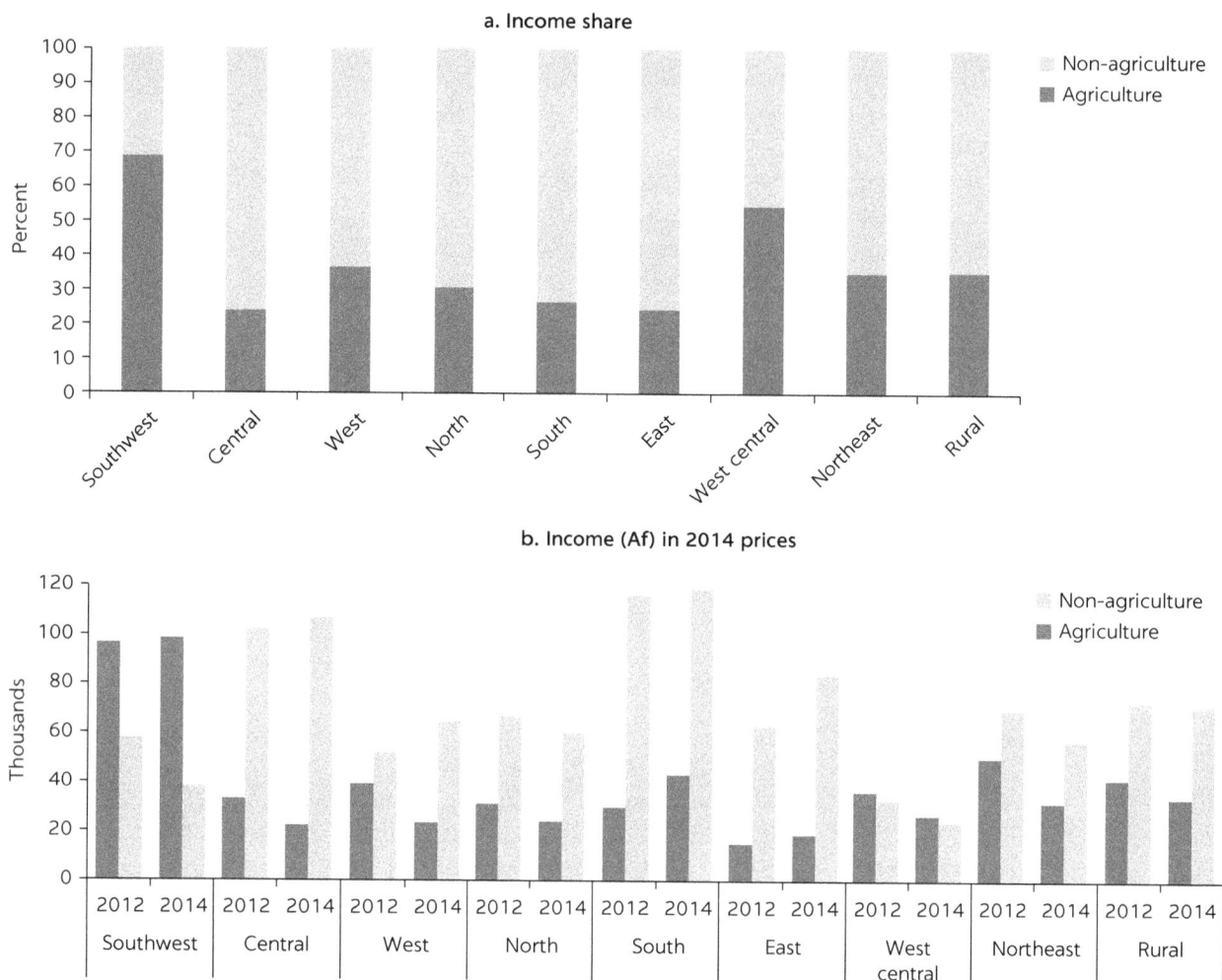

Source: Based on NRVA 2011–12 and ALCS 2013-14.

of nonfarm income in total income and the high concentration of nonfarm employment rates around Kabul suggest that proximity to urban areas matters in the sectoral composition of income and employment in rural areas (Deichmann, Shilpi, and Vakis 2008; Sen et al. 2014). In the southwest and west central regions, agriculture income share dominates household income.

The right panel in figure 2.10 reveals that agriculture income has gone up in the southwest, east, and south, but has fallen in all other regions. While agricultural income fell between 2012 and 2014, nonagricultural income remained stagnant, implying a gradual structural transformation of the rural economy toward the nonfarm sector in most regions.

Figure 2.11 presents employment type patterns across regions in 2014. The proportion of unpaid family workers among agricultural workers was high in most regions except the northeast. In the west central and central regions, for example, unpaid family workers accounted for 79 percent and 71 percent of farm employment, respectively. Unpaid family workers did not constitute a significant portion of nonagricultural workers in most regions, the west central region being

FIGURE 2.10

Spatial pattern of employment shares: regional level, 2013–14

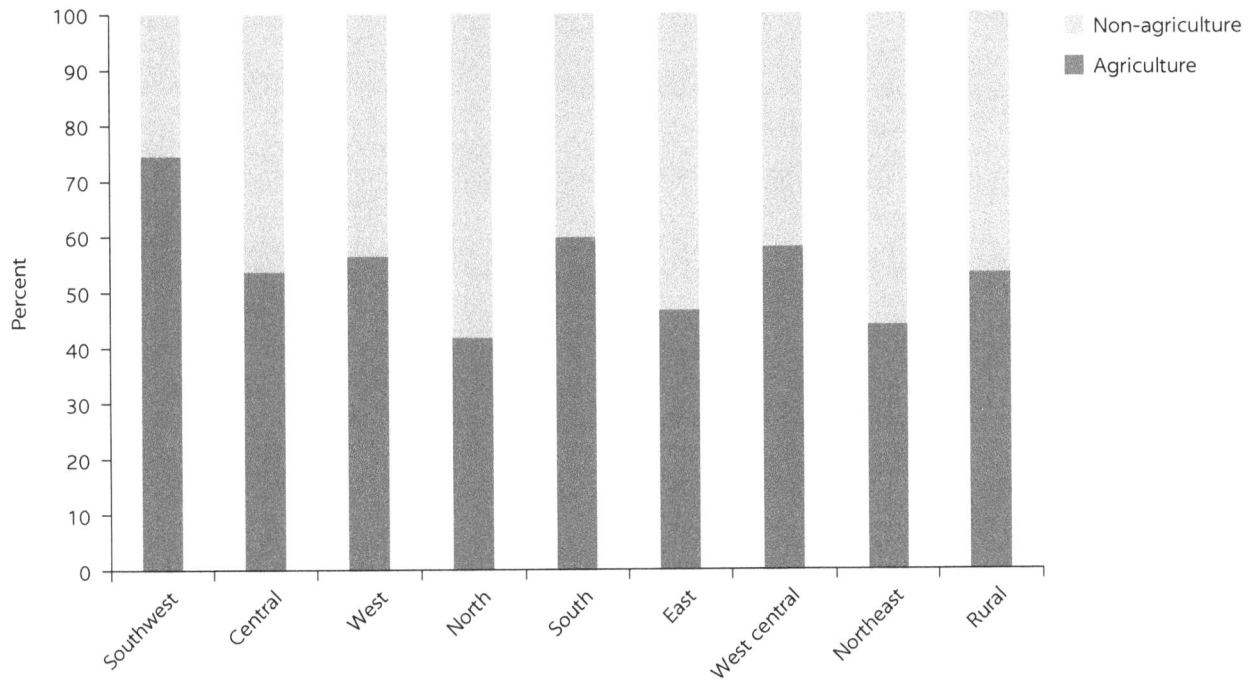

Source: Based on ALCS 2013–14.

FIGURE 2.11

Spatial pattern of employment types: regional level

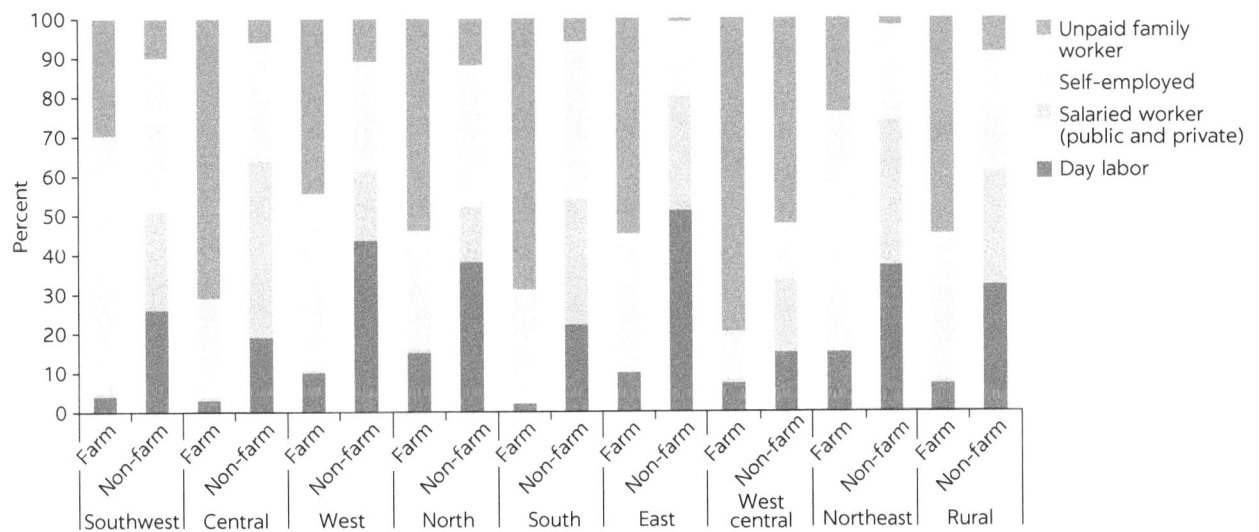

Source: Based on ALCS 2013–14.

one exception. The proportion of salaried workers in nonfarm employment was high in the central, south, east, and northeast.

The broad sectoral composition of rural Afghanistan suggests that the high share of agricultural employment is accompanied by a low share of income, with the low level of commercial farming and market participation and the rise of unpaid family workers. Moreover, the nonfarm sector's income share and employment share tend to be high in the central, south, and east regions around Kabul. The share of salaried workers in total employment is also high in these regions.

AGRICULTURE IN AFGHANISTAN: A VIABLE SECTOR FOR JOB CREATION?

The key factors in the low return from agricultural employment are the low level of market participation and the very high number of unpaid family workers in the sector, particularly among youth. Many questions, however, remain. What subsectors generate what type of employment? What causes a subsector to underperform or perform better? Is there any spatial variation in the job creation performances of subsectors? This section addresses these questions by analyzing primary survey data and secondary data.

Agriculture in Afghanistan: moving in what direction?

Agricultural labor has been increasing since the early 1990s (figure 2.12). During the same period, however, agricultural GDP remained stagnant, at around $3 billion.[2] Moreover, the consistent growth of the service sector decreased agriculture's share of total GDP. The increased share in agricultural employment, combined with a constant agricultural GDP, indicates a decline in agricultural labor productivity. Estimates of total factor productivity (TFP), a crucial measure of efficiency and thus an important indicator for policymakers, for agriculture demonstrate that agricultural productivity has always been erratic. While the TFP growth rate was about 1.5 percent in the first half of the 1990s, it became negative in the late 1990s and early 2000s. TFP growth increased to more than 2 percent in the late 2000s, and slowed again in early

FIGURE 2.12

Stock of agricultural labor in Afghanistan

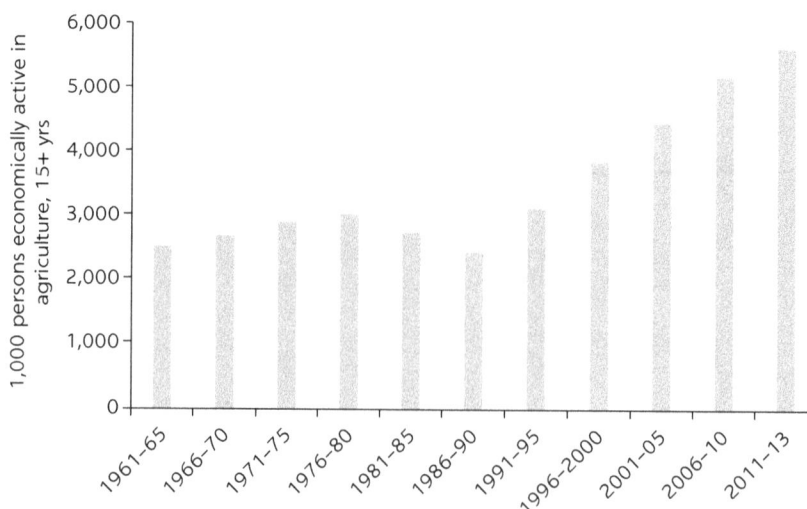

Source: U.S. Department of Agriculture 2016.

FIGURE 2.13

TFP, input, and output growth in agriculture

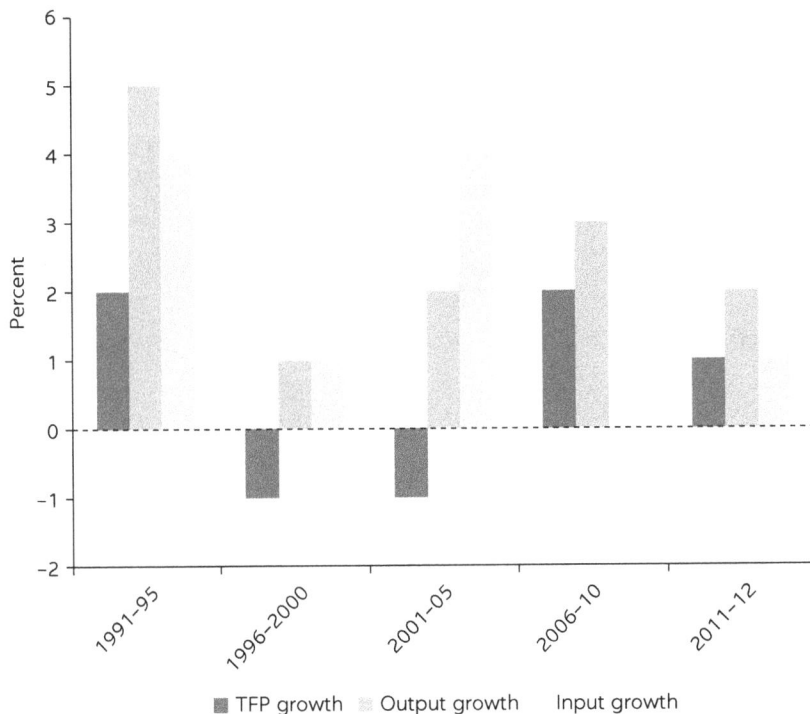

Source: U.S. Department of Agriculture 2016.

2010s; it increased by more than 2 percent in 2006–10, combined with a significant decline in input growth and a rise in output growth (figure 2.13). The lack of enough non-farm jobs in rural and urban areas could be the factor behind the simultaneous increase of agricultural labor-many of them are unpaid family labor-and agricultural TFP in Afghanistan.

In 2011–12, output growth declined compared with the late 2000s, while input growth increased, leading to a slowdown in overall TFP growth. Nonetheless, the 2 percent average TFP growth rate in the late 2000s implies that, given political stability and policy support, Afghanistan's agriculture sector has the potential for fast growth.

We examined the agriculture sector and its employment patterns to understand the jobs rural workers pursued (figure 2.14). In 2013–14, about 1.5 million of 2.5 million rural people employed in agriculture worked in the farm sector; the remaining 1 million worked in the livestock sector (figure 2.15). In the same period, about 1.3 million people in agriculture were unpaid family workers, the majority in the livestock sector. The livestock sector's share of total employment varied across regions. It was highest in the south, followed by the central, west, and north regions.

Information about income sources in rural areas allows a breakdown of farm income's share by major subsectors. Figure 2.16 presents income shares of agriculture subsectors in terms of total rural income, showing that shares from crop and non-crop agriculture fell in most rural areas, though they increased in the southwest, south, and east. Crop agriculture accounted for about 60 percent of the agricultural income of a rural household, while orchards, agricultural labor, and livestock also played a key role in income generation.

FIGURE 2.14

Employment types in agriculture

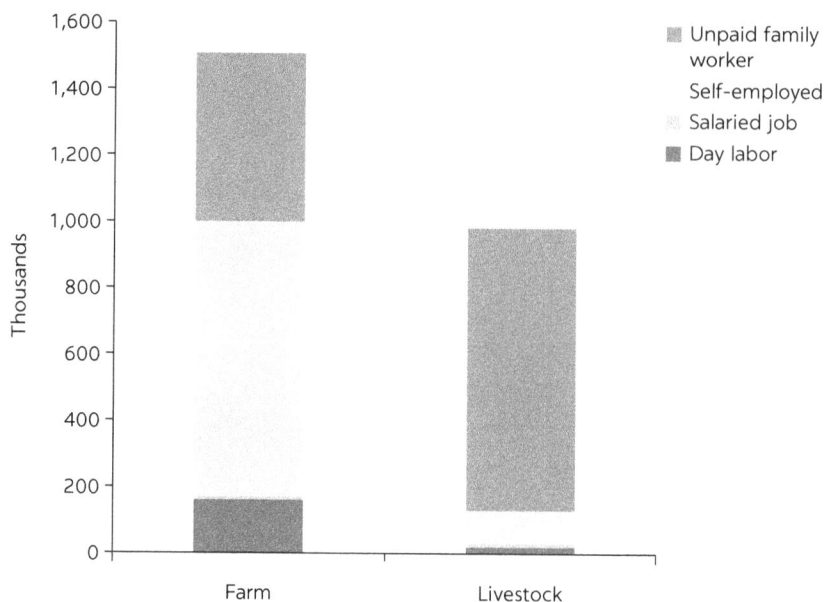

Source: Based on ALCS 2013-14.

FIGURE 2.15

Employment shares of agricultural subsectors: regional level, 2013-14

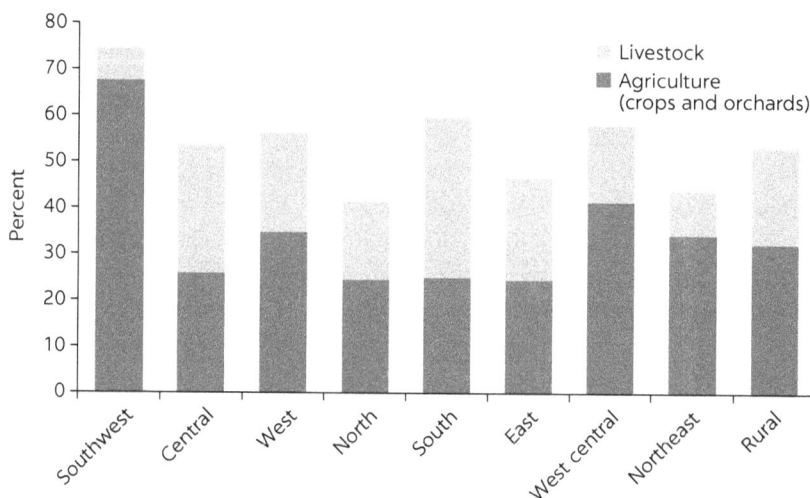

Source: Based on ALCS 2013-14.

The share of livestock in total income decreased in most regions, although it increased in the west central and northeast regions (figure 2.17). Income share from opium is insignificant except in the southwest and west, and shows a declining trend overall. In the southwest in 2011–12, opium's income share was around 8 percent, but fell to 5 percent 2013–14.

FIGURE 2.16
Income shares of agricultural sources, 2011–12 and 2013–14

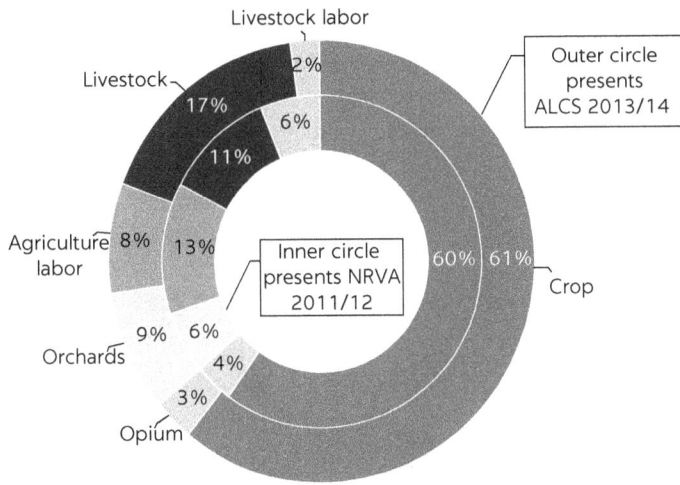

Source: Based on NRVA 2011–12 and ALCS 2013–14.

FIGURE 2.17
Regional annual household incomes from agricultural sources

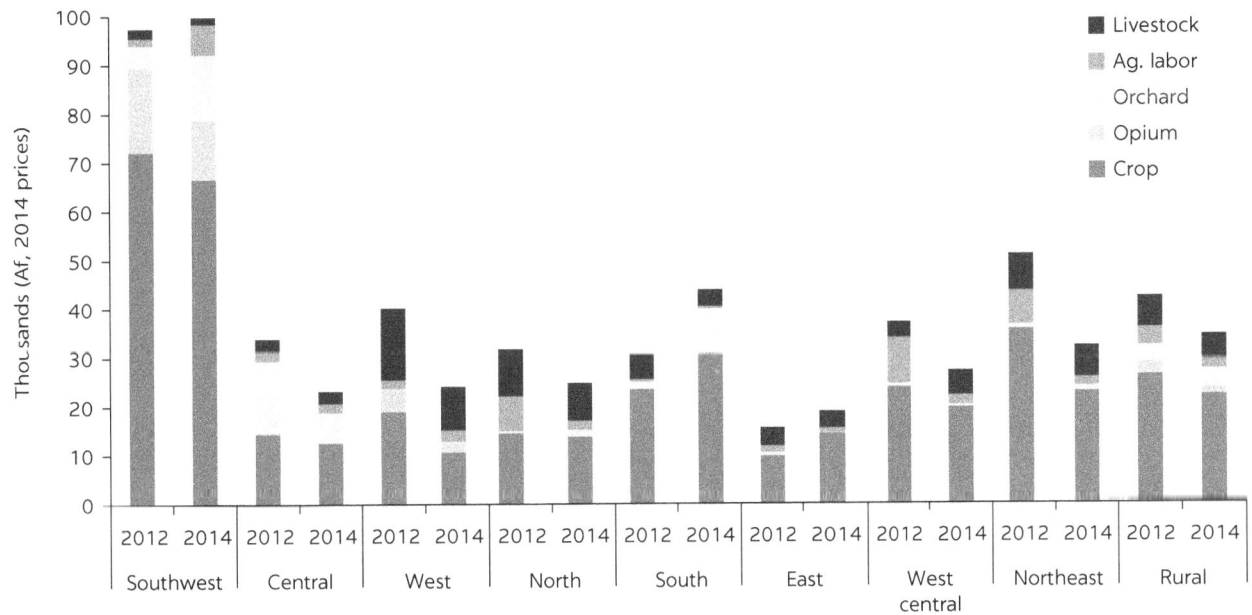

Source: Based on NRVA 2011–12 and ALCS 2013–14.

The livestock subsector generates about 40 percent of agricultural employment in rural areas. However, many livestock workers are unpaid family workers and very few of them participate in the market, resulting in increased earning per capita. The income share of the orchards subsector has seen positive changes, while most other subsectors, including farming and poppy cultivation, saw declines.

Crop agriculture: can a less diversified sector driven by food security create more jobs?

The income share of crop agriculture declined in most regions. This subsector is overly concentrated on wheat production, which uses about three-quarters of irrigated land (figure 2.18). Rice is the second most important crop grown on irrigated rural land, but accounts for only 6 percent of total irrigated land. The lack of diversification in crop agriculture has made farm households vulnerable to stagnant or declining wheat prices in local markets. Figure 2.19 shows that retail prices of wheat in major cities have remained

FIGURE 2.18

Distribution of cultivated land area, 2008–09 and 2015–16

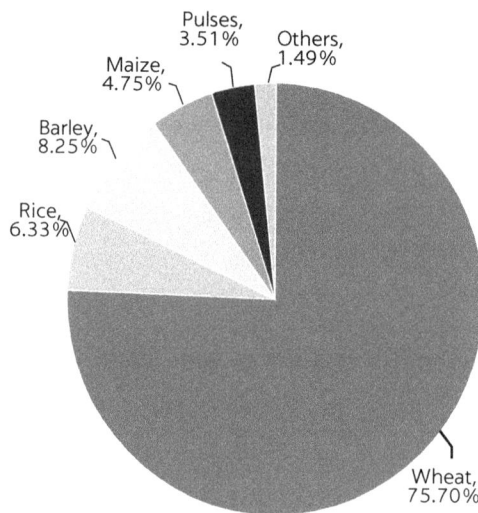

Source: Central Statistics Organization 2016.

FIGURE 2.19

Retail wheat prices in major cities, 2006-16

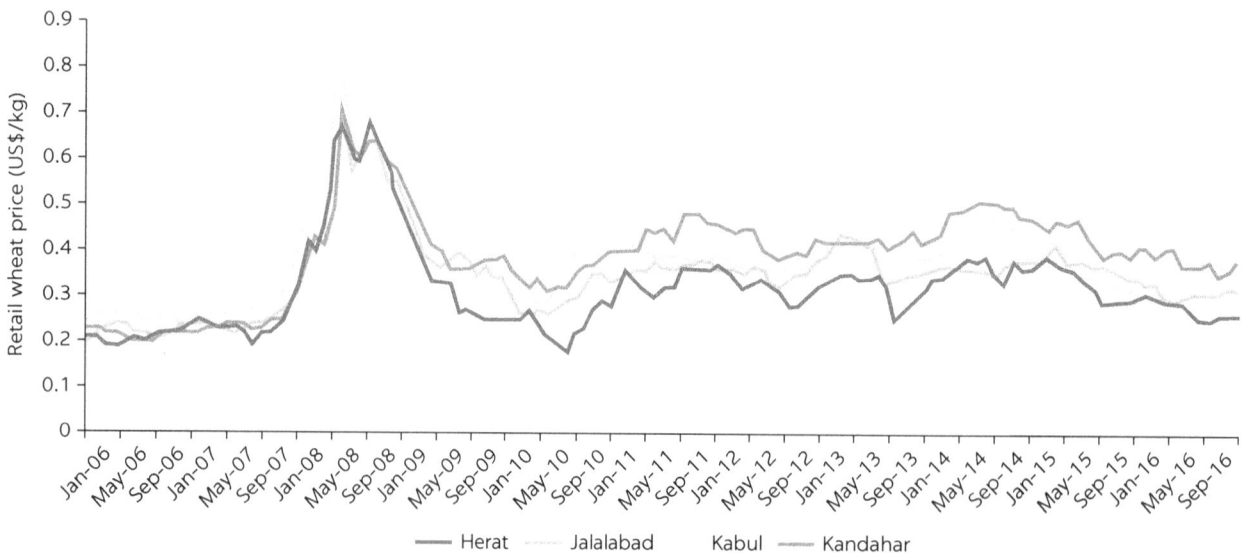

Sources: FAOSTAT, FAO, and Commodities Price Bulletin, Various Issues, MAIL, the government of Afghanistan.
Note: Prices were given in local currency in FAOSTAT, then converted into U.S. dollars using exchange rates of corresponding years.

FIGURE 2.20

Crop output growth by area, yield, and price, 2002–13

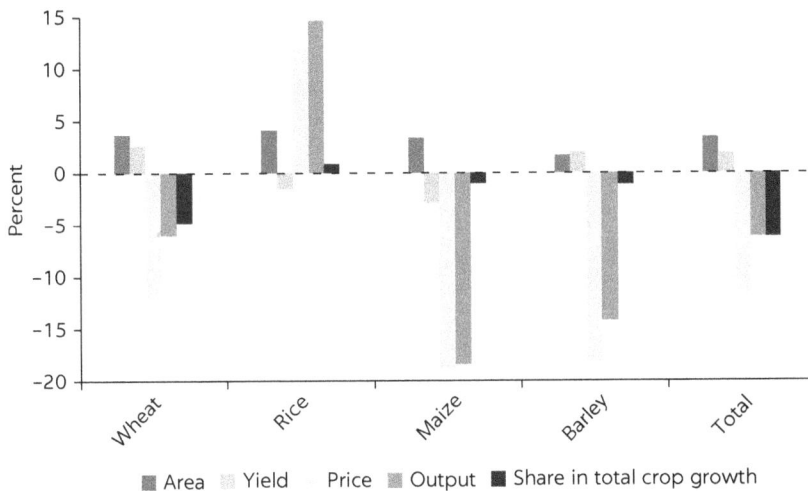

■ Area   Yield   Price ▨ Output ■ Share in total crop growth

Sources: Based on FAOSTAT production and price data.

stagnant since 2010, with a declining trend in recent years. Hence, the profitability of crop agriculture remained low.

To understand the sources of crop output growth, the change in output was divided into its three major components: area, yield, and price. This exercise uses data for 2002–13 on production quantities, area, yield, and for four major crops: wheat, rice, maize, and barley. The results are presented in figure 2.20, which shows that in the last decade the area under wheat production increased by 4 percent and the yield rate of wheat increased 3 percent. However, the consistent fall of wheat prices in domestic and international markets resulted in negative output growth. With so much irrigated land devoted to wheat cultivation, negative output growth affects most rural farmers. Furthermore, wheat's lack of profitability may prompt farmers to cultivate poppy on their irrigated land. This is why, in recent years, a gradual rise of area under poppy cultivation can be seen (UNODC 2015).

Crop agriculture faces another major constraint: irrigation. Some agricultural land lacks irrigation facilities and relies on rainfall for crop farming. The ratio of irrigated land to total agricultural land is high in the southwest and east (map 2.3), but low in the west and north. Agriculture in the west and west central regions is mostly rain-fed. But many households that own irrigated land have reported that they often lack sufficient irrigation (map 2A.1). Even if facilities exist, a lack of water may undermine irrigation's far-reaching benefits for productivity enhancement. For example, even though most agricultural land in Nimroz province is irrigated, farmers report insufficient quantities of water. Wheat remains the dominant crop in irrigated agriculture in most parts of the country, and only provinces in the west central and north regions exhibit some type of diversification in the use of irrigated land.

Employment in poppy/opium production

Poppy cultivation continues in rural Afghanistan despite numerous security and non-security measures to curb it. Sustained cultivation and the opium economy are often attributed to low profitability and returns from crop agriculture, and

MAP 2.3

Proportion of irrigated land in total cultivated land, 2013–14

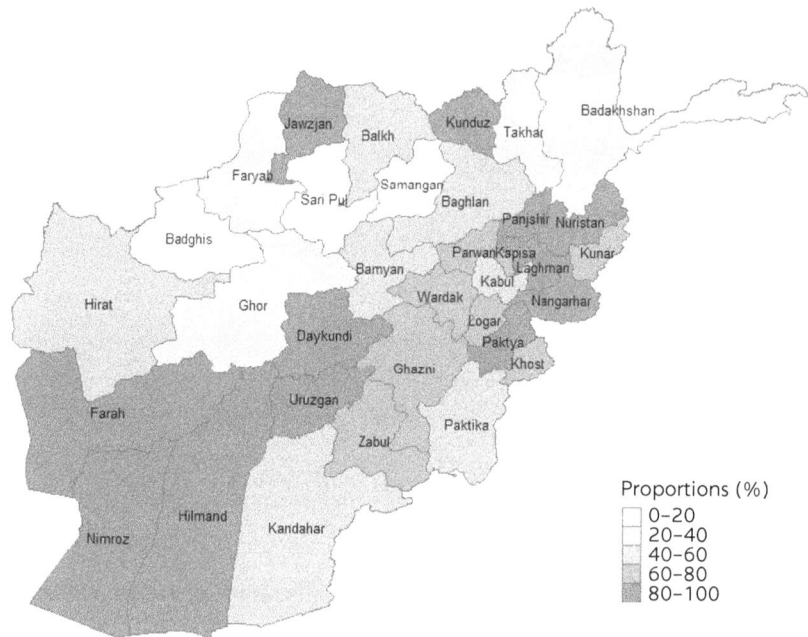

Proportions (%)
- 0–20
- 20–40
- 40–60
- 60–80
- 80–100

Source: Based on ALCS 2013–14.

few opportunities for off-farm employment. Difficult market access for staple crops also plays a key role. The United Nations Office on Drugs and Crime (UNODC 2015) reported that, in 2015, farmers who stopped poppy cultivation and switched to wheat production earned less annual income per household than those who continued poppy cultivation ($2,946 vs. $3,434). Poppy is a cash crop; in a fragile security context, farmers can get involved in cultivation to simply earn cash quickly to buy wheat flour, rice, and other staples.

Although less than 3 percent of total cultivable land is used to grow poppy, opium remains Afghanistan's most important cash crop and its largest export item, with an estimated value of $3 billion in border prices (Byrd and Mansfield 2014). Afghan exports account for 90 percent of the global supply of opiates (SIGAR 2014). The opium economy also has significant employment effects, generating about 411,000 full-time equivalent jobs directly, and supports indirect job creation in the illicit economy (Byrd and Mansfield 2014). Estimates suggest that revenues from illegal opium trade are equivalent to one-third of Afghanistan's reported GDP (Ward et al. 2008). Furthermore, the area under poppy cultivation grew from 57,000 hectares in 2007 to 224,000 hectares in 2014 before declining to 183,000 hectares in 2015 (UNODC 2015).

With support from donor countries, the government of Afghanistan has been implementing strategies to curb opium production. Poppy cultivation, however, remains popular among a few groups of farmers in rural areas. Without viable alternatives, it has proven difficult to reduce opium production solely with law enforcement measures (Ward et al. 2008), which often fail to address the structural and institutional aspects that contribute to the growth of poppy cultivation (Mansfield and Pain 2007). Furthermore, it is difficult to curb poppy cultivation without making wheat production a viable alternative. Cultivating poppy is three times more profitable than growing wheat (figure 2.21); consequently, it is less attractive to grow wheat, and the government's strategies to curb poppy production remain less successful.

There are several reasons why the size of the area under poppy cultivation does not always follow the direction of opium prices, and why the amount of land being used to grow poppy increased while opium prices dropped in 2004–09 and 2012–14. Figures 2.22 and 2.23 present two series: the index of ratios between opium prices and wheat prices, and the index of ratios between the area under poppy cultivation and wheat cultivation.

Although the price of opium declined between 2004 and 2009, its relative price increased due to falling wheat prices. This motivated farmers to increase their land area under poppy production. Since 2011, however,

FIGURE 2.21

Gross income per hectare from poppy and wheat cultivation

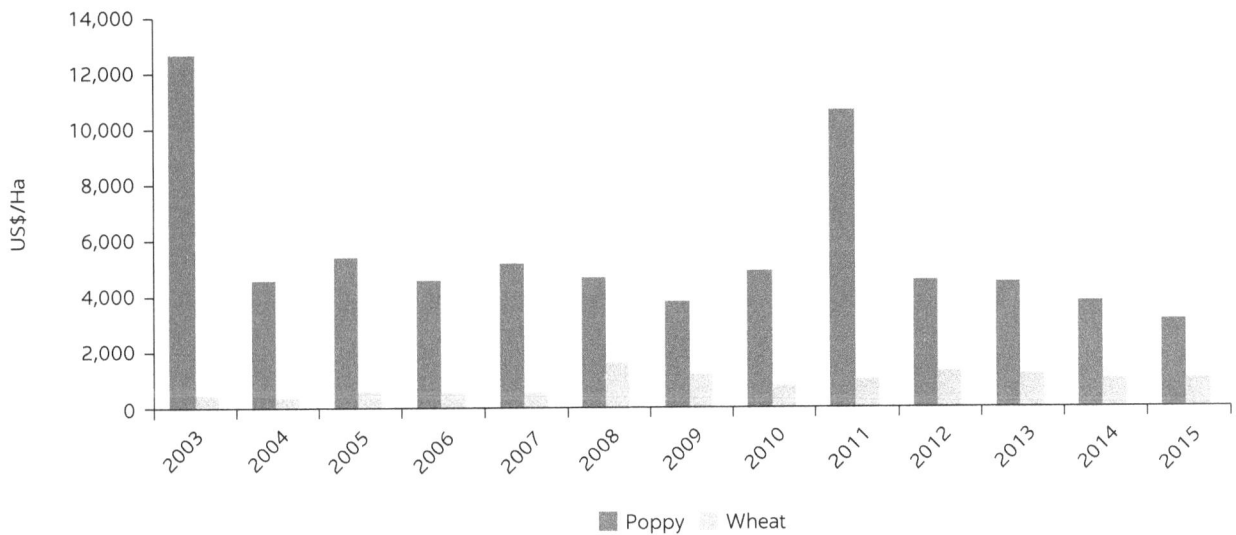

Source: Adapted from UNODC 2015.

FIGURE 2.22

Cultivation area and price of opium

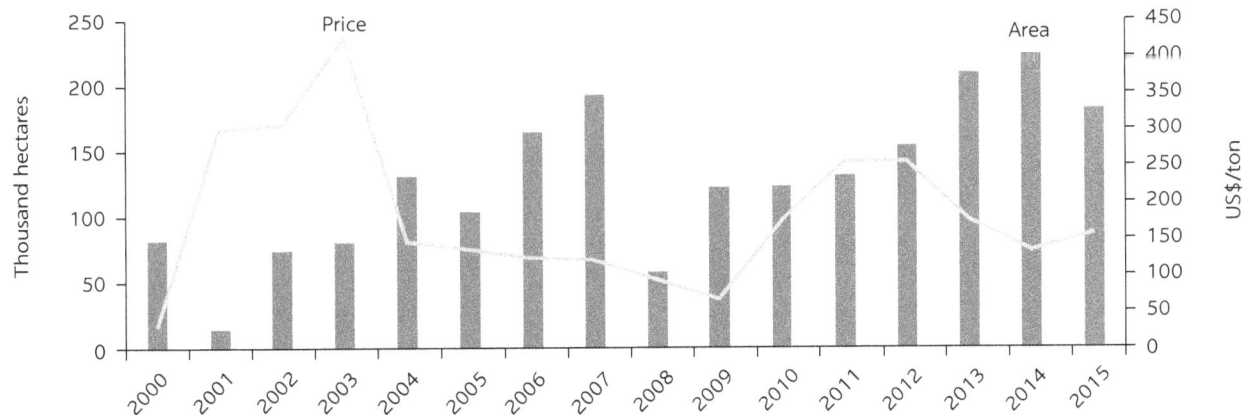

Source: Based on data from UNODC 2015.

FIGURE 2.23

Interplay between wheat and opium production

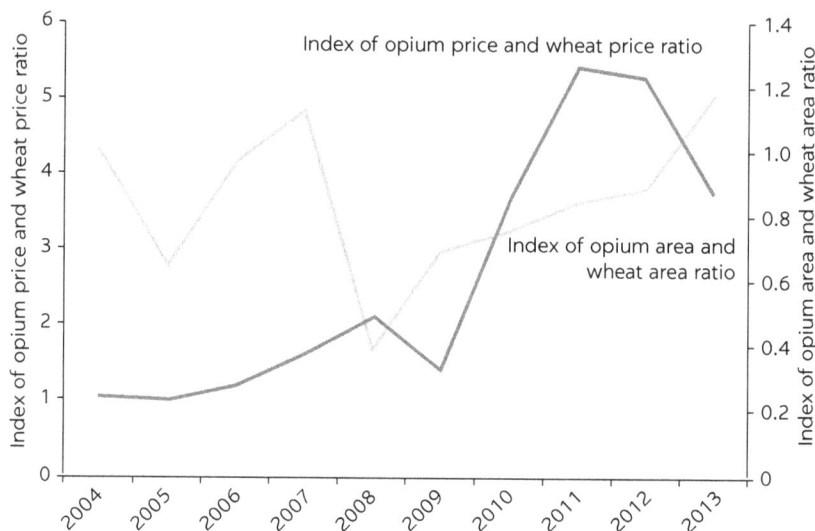

Sources: Based on data from UNODC 2015 and FAOSTAT 2016.

the area under poppy production increased as the relative price of opium decreased. This suggests the decision to cultivate poppy is motivated by more than the relative price of opium—it is motivated by the overall security situation and presence of the Taliban. The success of opium production reduction strategies depends on measures such as developing viable alternatives to wheat, increasing wheat profitability, and encouraging crop diversification. They also depend on sustainable security measures by addressing the structural causes of rural poverty, such as the lack of physical and social infrastructure.

Horticulture economy: a sector with growing employment opportunities

Historically, horticulture has played a significant role in rural livelihoods in Afghanistan. Today, fruits and nuts produced in Afghanistan are in great demand and have increasing potential for job creation across the value chain. However, it is not possible to estimate the number of people engaged in horticulture in rural areas with the available data. In 2013–14, the income share from orchards in the south and southwest increased, but declined in the central region. Ownership of garden plots and earning income from orchards are concentrated in the central and southwest regions, home to Kabul and Kandahar (figure 2.24).

Figure 2.25 shows that the area under noncrop agriculture has increased since 2014. Figure 2.26 shows that producer prices of major horticulture products were stagnant in the early 2010s.

About 16 percent of rural households have garden plots, but only 6 percent have orchard income, implying that they are using plots for their own consumption, not for commercial production of high-value fruits and nuts. In the north, south, and west central regions, most rural households with garden plots do not earn any income from orchards. With the provision of technical and financial support to these households and better access to market facilities,

FIGURE 2.24

Rural households that own garden plots and receive orchard income, 2013–14

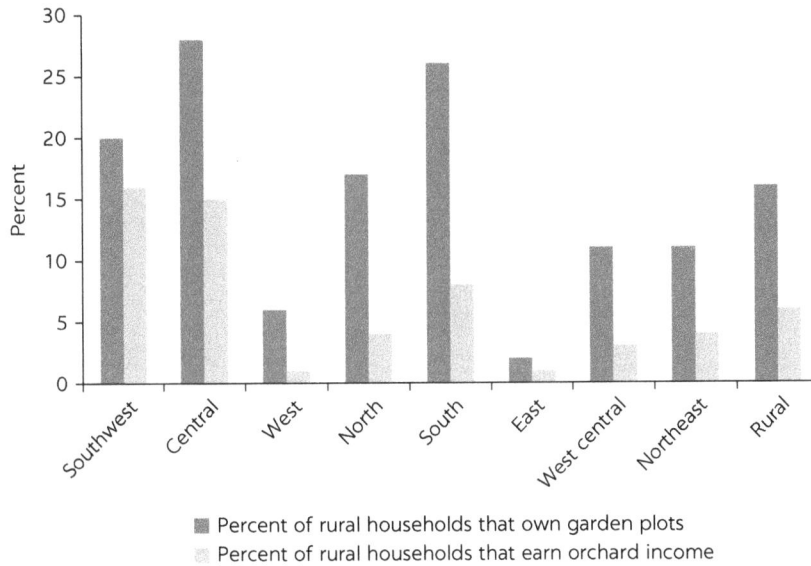

Percent of rural households that own garden plots
Percent of rural households that earn orchard income

Source: Based on ALCS 2013–14.

FIGURE 2.25

Area under different high-value noncrop production

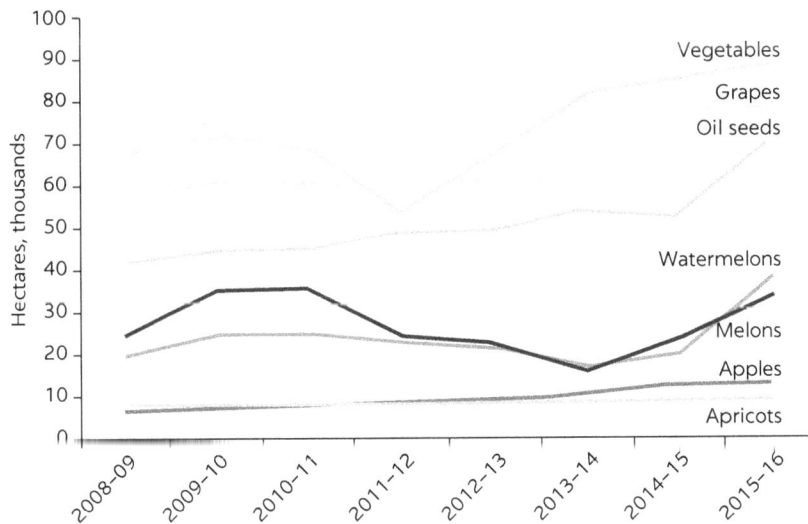

Source: Central Statistics Organization 2016.

the horticulture sector has significant potential for rural job creation. Commercial production of fruits and nuts in garden plots would not only raise the income and employment of garden owners, it would help create new jobs for young workers across the fruit and vegetable value chains. Accordingly, by improving the horticulture economy, the government could also raise the food processing sector's share of employment.

FIGURE 2.26

Producer price indices

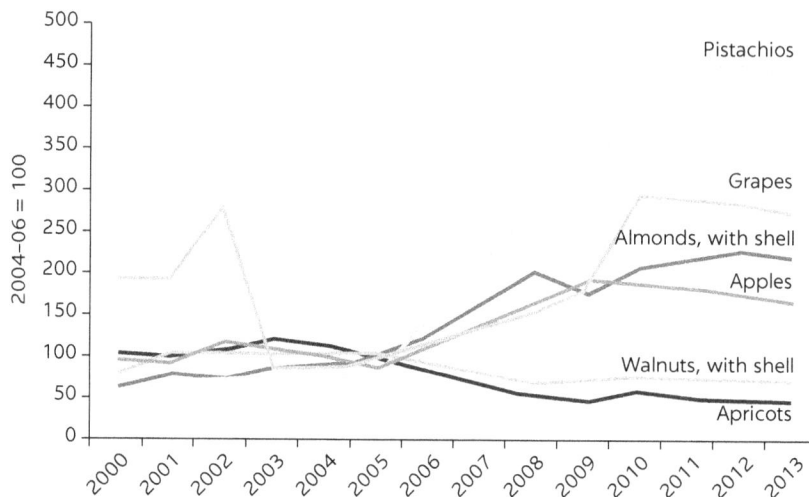

Source: Based on data from FAOSTAT 2016.

FIGURE 2.27

Livestock capital in rural Afghanistan

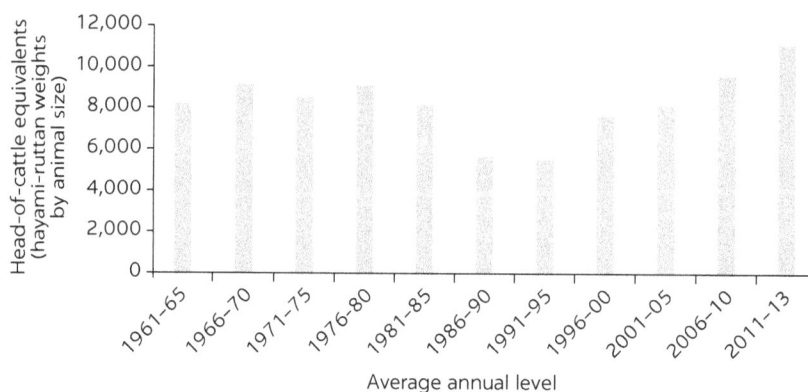

Source: U.S. Department of Agriculture 2016.

Livestock sector: a "sponge" for unpaid family workers?

The livestock sector is strategically important for promoting rural growth and creating rural jobs. Productivity growth and employment generation depend on increasing the productivity of small-scale producers, promoting commercial production, providing extension services, and strengthening market linkages for small and poor livestock producers.

The sector accounts for a large share of agricultural employment in rural areas, though most workers are unpaid family workers and very few households sell their livestock products in markets. Most rural households rear livestock, and about one-fifth of employed rural workers are engaged in livestock. Overall livestock capital has increased in recent years, but the sector accounts for only 6 percent of rural households' income, implying a low return to labor (figure 2.27).

Two factors may be responsible for the low employment return in the sector: the "youth bulge" and low market participation. Because there are insufficient paid work opportunities in nonfarm sectors, youth workers are increasingly joining in their families' agricultural activities as unpaid family workers, especially in the livestock sector. Also, few households sell their livestock products in the market, which implies that most are rearing livestock for their own consumption, not to generate income. The lower half of map 2.4 shows that, in most provinces, livestock's income share is much lower than its employment share. Thus, there is considerable unrealized potential to increase income by selling livestock. Connecting rural livestock producers with national value chains of livestock products is key to harnessing this potential.

MAP 2.4

Livestock owned and market participation across rural Afghanistan, 2013–14

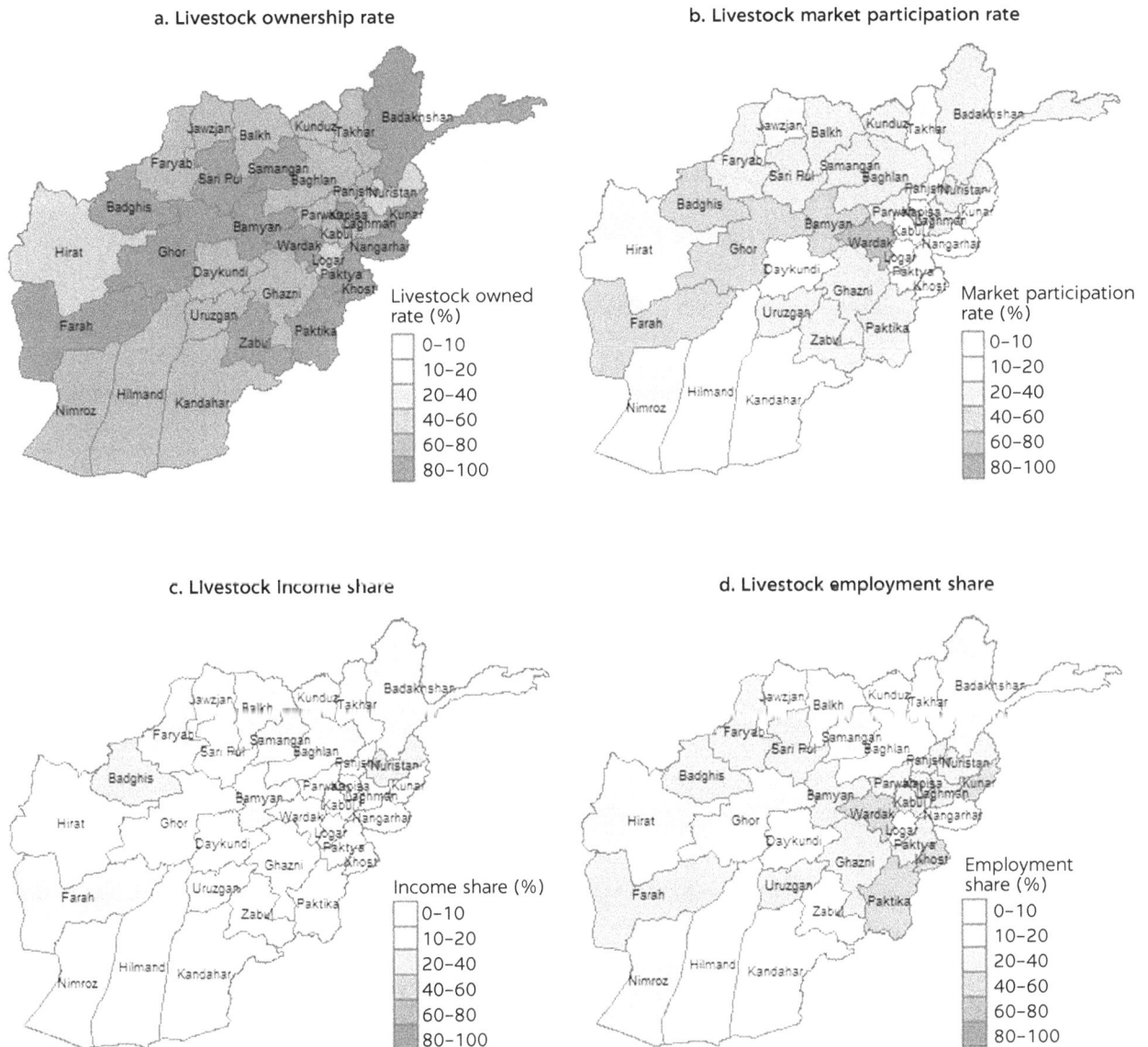

a. Livestock ownership rate

b. Livestock market participation rate

c. Livestock income share

d. Livestock employment share

Source: Based on ALCS 2013–14.

Farm and nonfarm linkages

As important as the farm and nonfarm sectors of the rural economy are, it is vital to recognize their interdependence and critical roles in accelerated and robust structural transformation. The prosperity of the rural nonagricultural economy largely depends on rural agriculture's performance. The importance of agricultural linkages in driving development, especially in the rural nonfarm economy, has long been recognized among development economists and practitioners. Many rural nonagricultural activities are strongly linked to agriculture, such as food processing, trading, and food preparation.

Through strong forward and backward linkages, any increase in agricultural income generates an increase in nonagricultural incomes in rural areas. For example, many studies in the development literature report a positive association between agricultural productivity and nonfarm employment (see, for example, Lanjouw and Lanjouw [2001], and Foster and Rosenzweig [2004]). Using data from Bangladesh, Shilpi and Emran (2015) estimate a significant positive impact on agricultural productivity on the growth of informal manufacturing and skills services employment. Gautam and Faruqee (2016) have also shown that agriculture is a major driver of rural nonfarm growth in Bangladesh. Last, agricultural productivity also supports nonfarm job creation in rural areas by increasing demand for goods and services.

In rural Afghanistan, many nonfarm activities are still linked to agriculture, consistent with the transformation process. And while some are "pull" jobs created to supply inputs for agricultural production or processing primary products, others are "push" jobs, generally those that small farmers or the landless are forced into out of necessity. Thus, policies to improve the productivity and income of people engaged in agriculture could complement the policies for job creation in the rural nonfarm sector.

Employment patterns and dynamics in the nonfarm sector

Available data indicate that rural nonfarm activities are diverse, including agro-processing, commercial/service activities, construction, manufacturing, trading, and transportation. As noted above, though activities in the nonfarm sector play a key role in drawing rural income, they have less employment, implying a higher labor productivity. Higher incomes in nonagricultural activities are mainly driven by service sector workers. Figure 2.28 presents employment patterns in the nonfarm sector. About one-third of the rural nonfarm workforce (655,000 workers) is employed in the construction sector; of those, more than 80 percent (553,000) are wage laborers. Therefore, construction activities, which are mainly supported by donor communities through their efforts to rebuild Afghanistan, do generate employment, but not decent and sustainable employment.

The second critical nonfarm sector is the formal services sector, which includes education, health, and other public services, accounting for about one-fifth of rural nonfarm employment (439,000 workers). Most jobs in this sector are salaried, provide benefits, and are sustainable in terms of job tenure.

Manufacturing and trade are the other major subsectors in the rural nonfarm sector, accounting for about one of every three jobs. While trade jobs are overwhelmingly done through self-employment in small and medium-sized enterprises, manufacturing jobs include many unpaid family workers. People employed in other services are also overwhelmingly in salaried jobs (figure 2.29).

FIGURE 2.28
Employment in the nonfarm sector, 2013–14

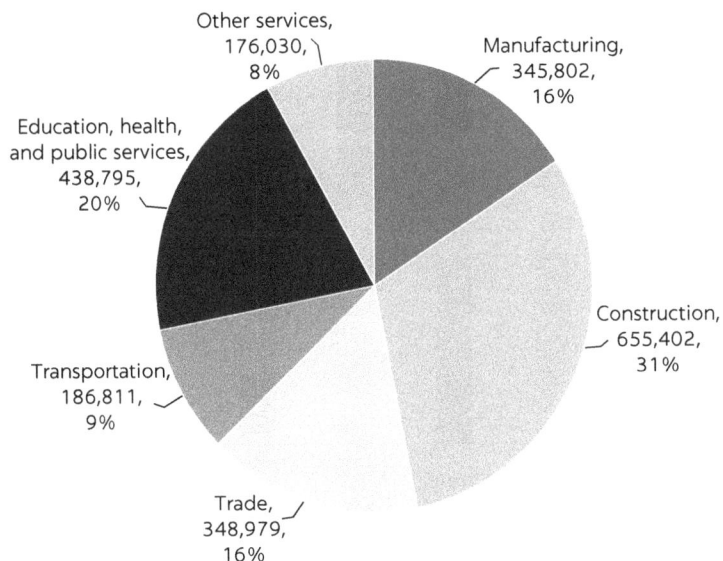

Source: Based on ALCS 2013–14.

FIGURE 2.29
Employment dynamics and types in nonfarm agriculture

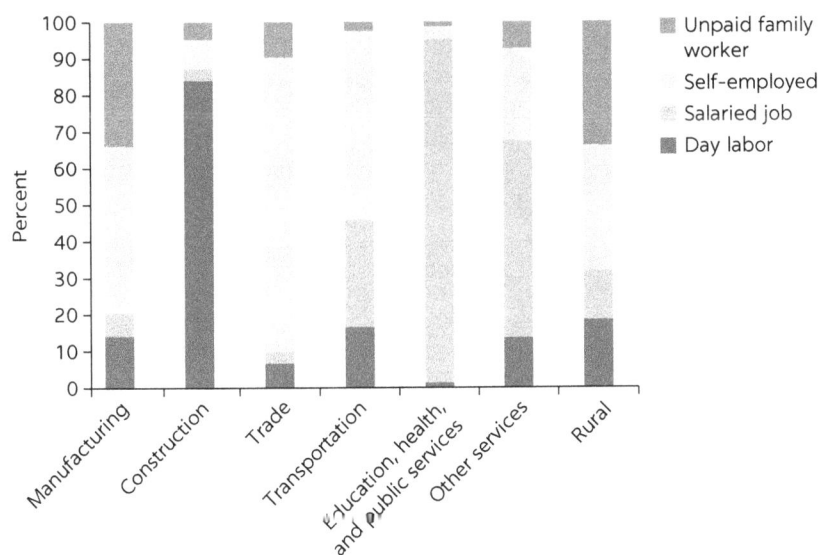

Source: Based on ALCS 2013–14.

As shown in figure 2.30, the composition of nonagricultural activities varies greatly across regions. Construction accounts for a significant portion of nonagricultural rural employment in most regions, carrying more weight than other nonfarm activities in terms of employment generation. Manufacturing is important for employment generation in the west central and north regions, while wholesale and retail trade contribute more in the southwest, south, and central regions. Salaried workers in health, education, NGOs, and other public services are a substantial part of nonfarm employment in the central, east, and northeast regions. The employment share of the transportation sector is small, however, with about

FIGURE 2.30

Employment shares in nonagricultural activities: regional level

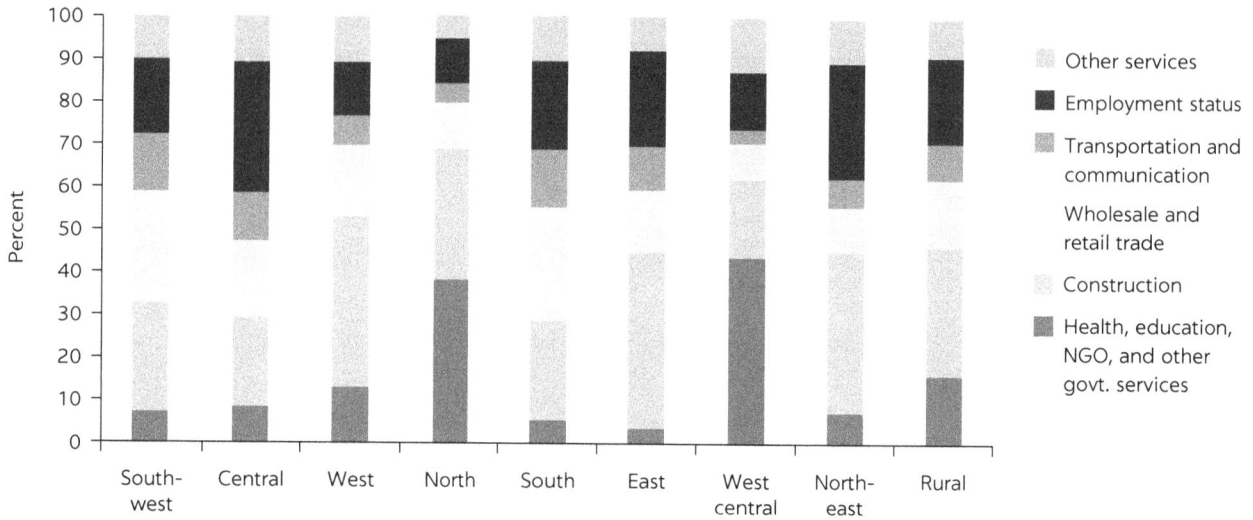

Source: Based on ALCS 2013–14.

Note: NGO = nongovernmental organization.

FIGURE 2.31

Income share of nonagricultural sources

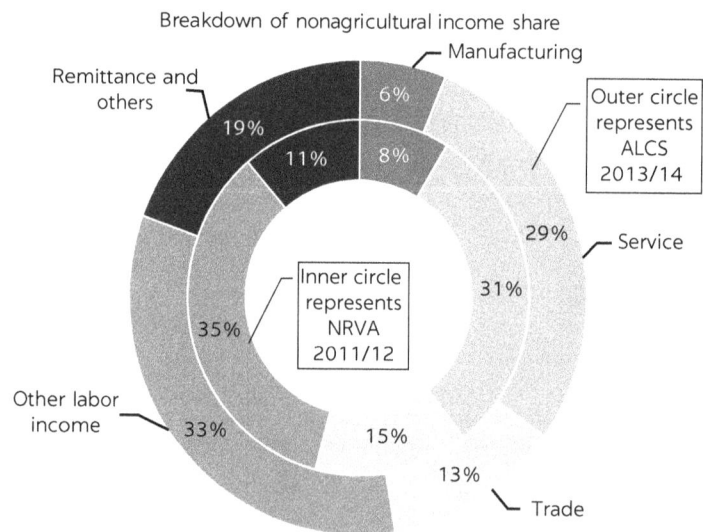

Source: Based on NRVA 2011–12 and ALCS 2013–14.

4 percent of rural workers; its share is above 5 percent only in the central, south, and east regions around Kabul. This clustering of transportation employment in rural areas close to the capital leads to two conclusions: proximity to urban areas affects employment, and because the overall rural transportation infrastructure is weak, the sector employs fewer workers.

In rural areas, the share of nonfarm income increased between 2011 and 2014, primarily driven by remittances and other transfers, which increased from 11 percent in 2011–12 to 19 percent in 2013–14 (figure 2.31). About 20 percent of rural nonfarm workers are involved in the education, health, and public services sector, earning about 29 percent of all rural income. This implies a higher return in salaried work, which often requires a better set of skills and training.

Figure 2.31 also shows that trade and manufacturing incomes accounted for about 20 percent of nonfarm income in 2013–14. Though manufacturing's employment share has increased, its income share and absolute income has dropped across all regions. Further analysis is needed to determine what could be driving this decline, though the high number of unpaid family workers in the sector may be a factor.

Absolute household income from remittances and other transfers also increased, from Af 6,837 in 2011–12, to Af 12,994 in 2013–14 (figure 2.32).

FIGURE 2.32

Regional income share in the nonfarm sector

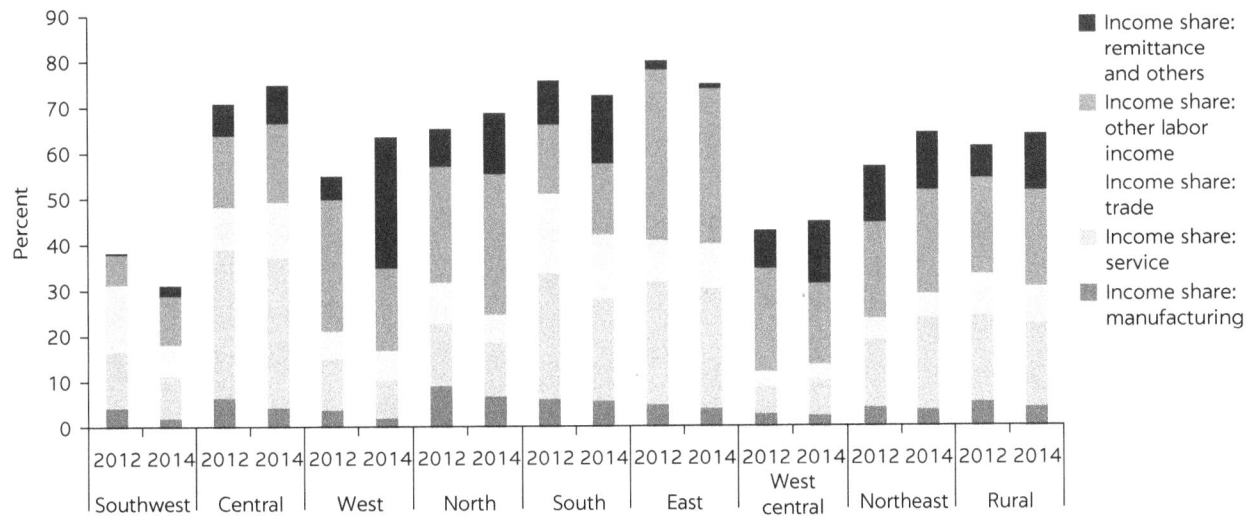

Source: Based on NRVA 2011–12 and ALCS 2013–14.

FIGURE 2.33

Regional household income from non-agriculture sources

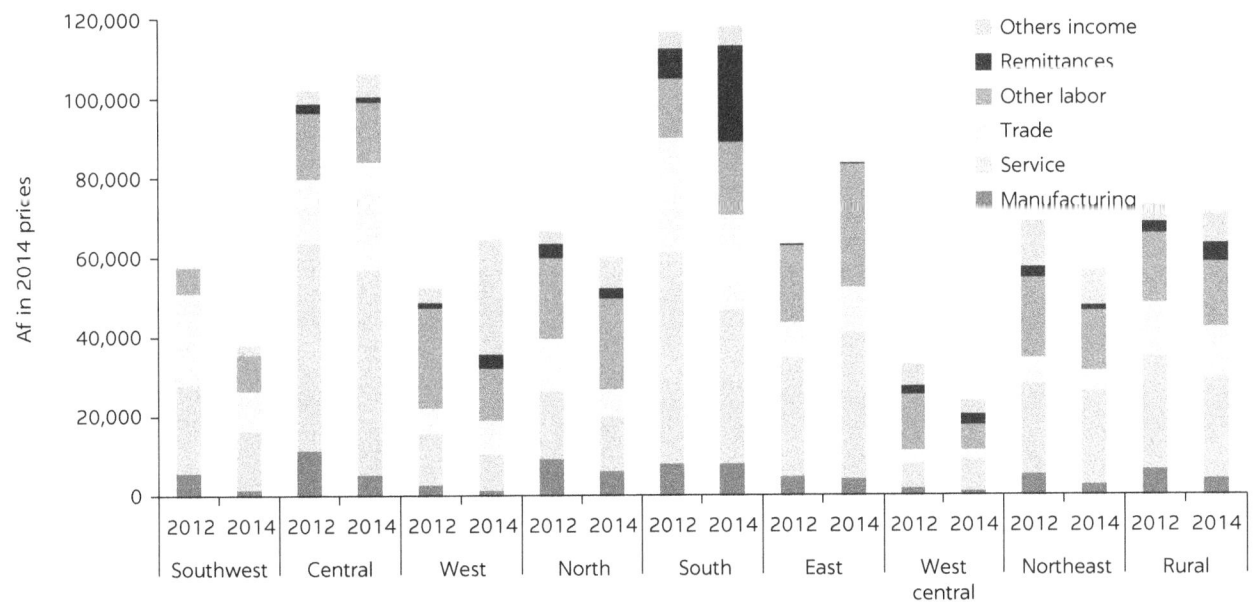

Source: Based on NRVA 2011–12 and ALCS 2013–14.

Wage labor is the most important nonagricultural income source, followed by the service sector. The share of income from different sectors varies widely. Rural households around Kabul (for example, in the central, east, and south regions) earn more than 70 percent of their total income from the nonfarm sector. In most regions, the share of remittances and other transfers has increased, while shares of total income and absolute income in all other non-agriculture sources have declined or stagnated.

While the overall share and absolute income from trade and retail services have declined, both increased in some regions (for example, the central and east regions) (figure 2.33). However, the share of income declined in the southwest and south.

MAP 2.5

Spatial distribution of employment share: construction and manufacturing, 2013–14

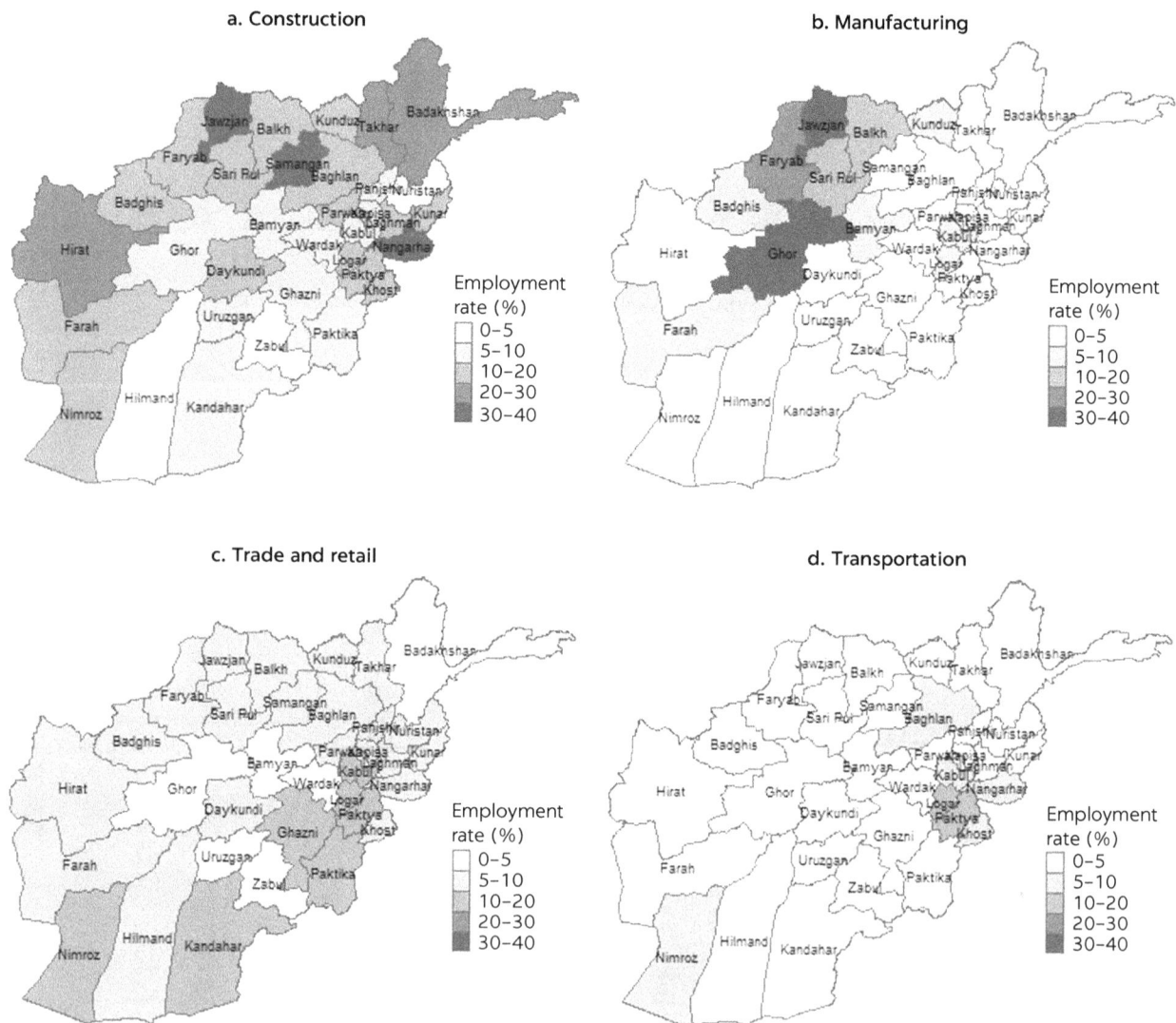

a. Construction

b. Manufacturing

c. Trade and retail

d. Transportation

Source: Based on ALCS 2013–14.

In 2011–12 and 2013–14, the income share of services fell or remained stagnant in all but the west central and northeast regions, which had increases of 5 percent and 2 percent, respectively. The share of other labor income as part of total income increased considerably in the southwest and the north, but declined significantly in the west.

The employment share in different nonfarm sectors varies across provinces (map 2.5). The share of the construction sector, where many workers are engaged in donor-funded road construction activities, is more than 10 percent in most provinces. The intensity of road construction has increased considerably since 2010 (figure 2A.2). More people are working in the construction sector in provinces in the north, northeast, and west, as well as in the provinces around Kabul. In Helmand, Zabul, and Nuristan, however, the sector's employment share is less than 5 percent.

Manufacturing's share of employment in most provinces is less than 5 percent. Some rural manufacturing jobs are clustered in the northern provinces, and it is surprising that there are not more jobs in the provinces around Kabul (less than 10 percent in Kabul and Logar, and less than 5 percent in most provinces adjacent to Kabul). Employment in the trade and retail services is spread across rural areas; in most provinces, 5–10 percent of people are involved in these sectors. Transportation-related employment is centered in the provinces around Kabul.

Annex 2A

FIGURE 2A.1

Population age structure in rural Afghanistan

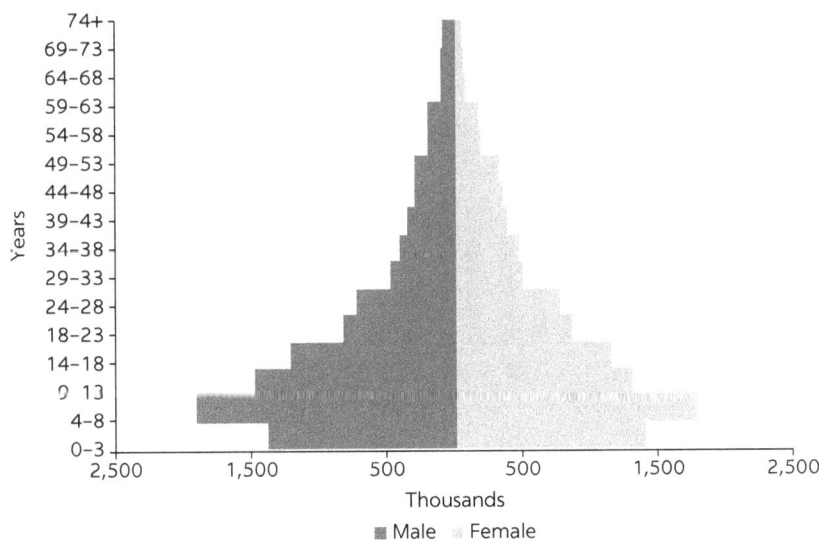

Source: Based on ALCS 2013–14.

FIGURE 2A.2
Total constructed roads

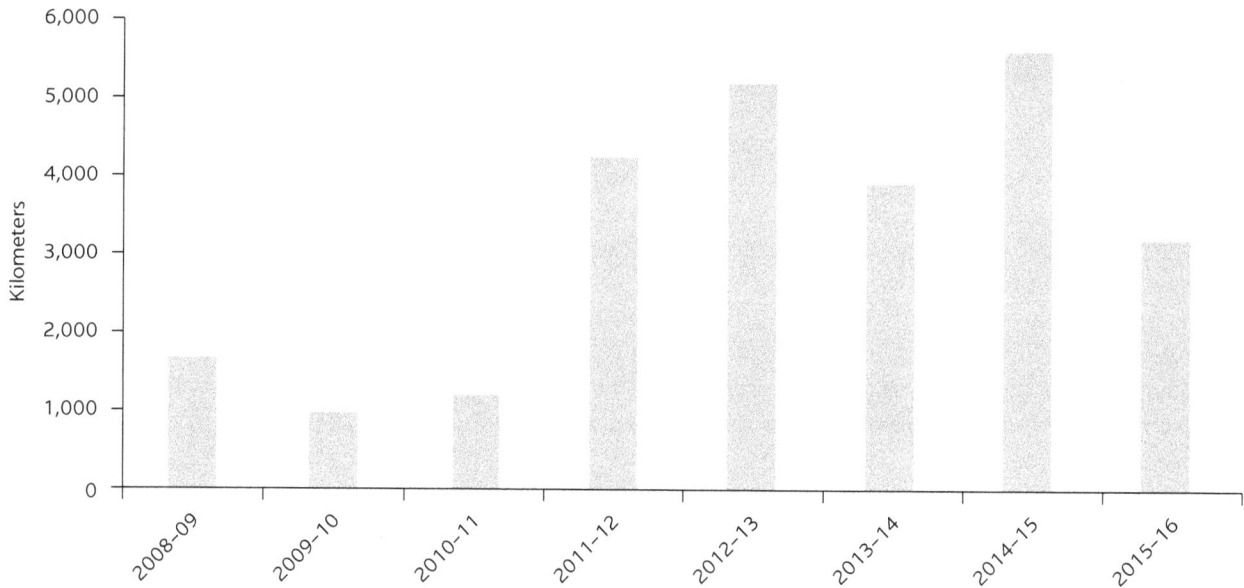

Source: Central Statistics Organization 2016.

FIGURE 2A.3
Annual per capita income in rural provinces, 2013–14

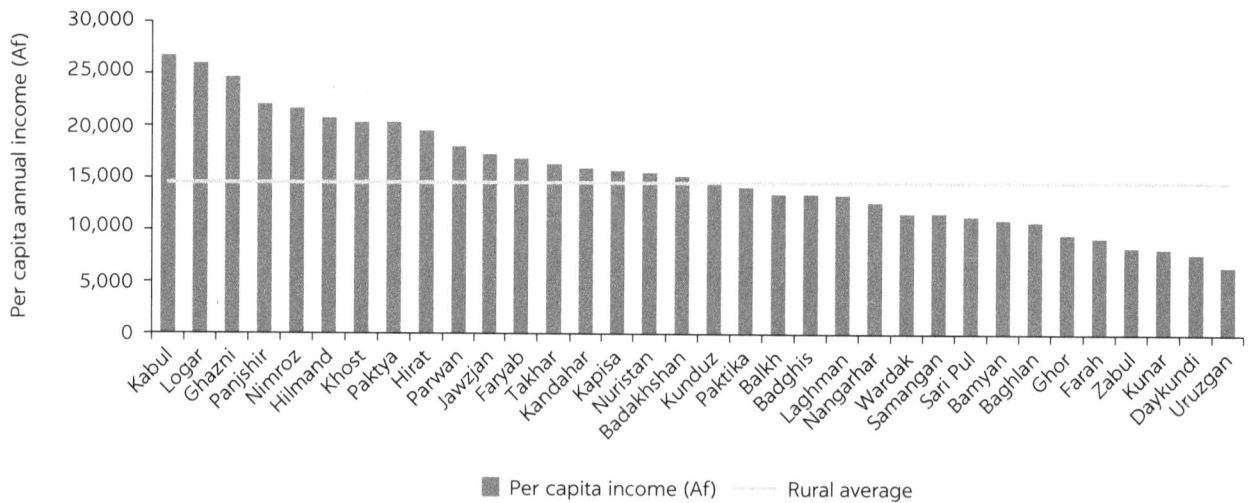

Source: Based on ALCS 2013–14.

MAP 2A.1

Households with irrigated land reporting lack of sufficient irrigation, 2013–14

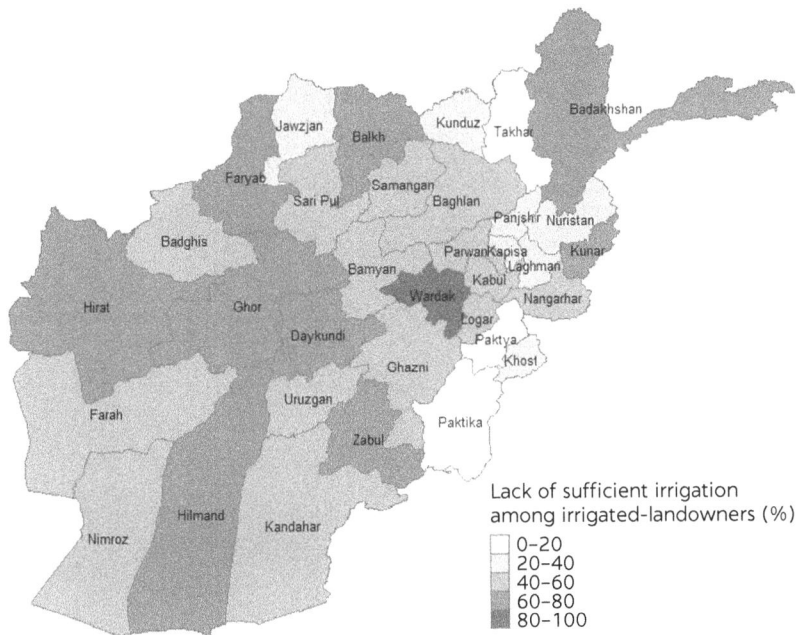

Lack of sufficient irrigation among irrigated-landowners (%)
- 0–20
- 20–40
- 40–60
- 60–80
- 80–100

Source: Based on ALCS 2013–14.

MAP 2A.2

Wheat as most important crop among irrigated-landowners, 2013–14

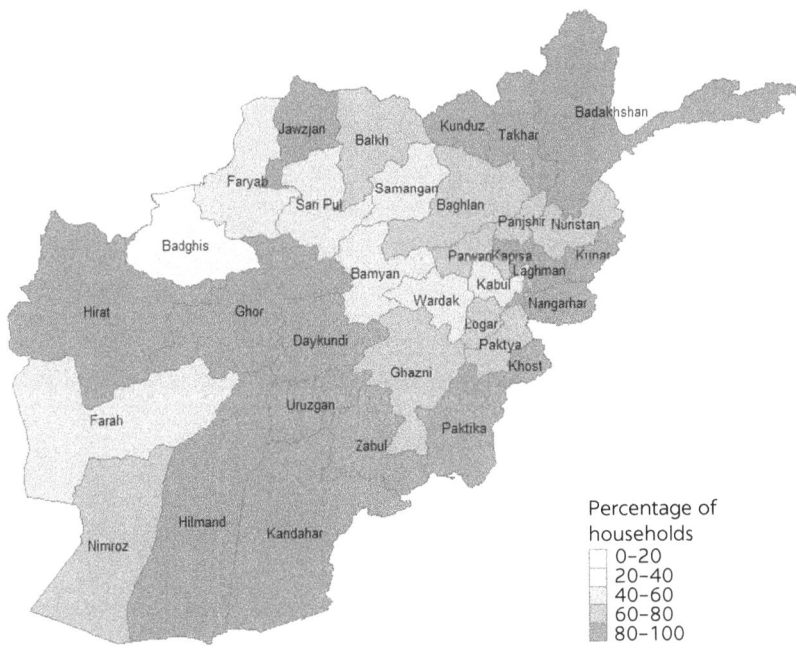

Percentage of households
- 0–20
- 20–40
- 40–60
- 60–80
- 80–100

Source: Based on NRVA 2011–12.

MAP 2A.3

Percent of rural households that own garden plots and receive orchard income

a. Garden plot owner

b. Market participants

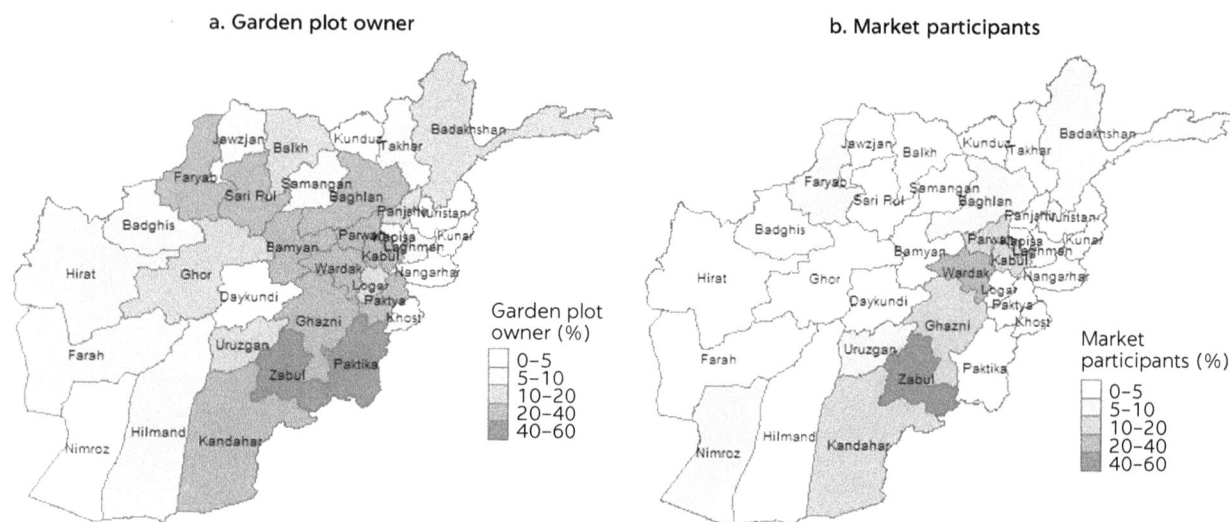

Source: Based on the ALCS 2013–14.

NOTES

1. All 2011–12 income statistics are deflated with the consumer price index of Afghanistan from World Development Indicators data.
2. Agricultural GDP was $2.9 billion in 1995 and $3.0 billion in 2012 at 2005 market prices (World Development Indicators 2016).

REFERENCES

Byrd, W., and D. Mansfield. 2014. "Afghanistan's Opium Economy: An Agriculture, Livelihoods and Governance Perspective." A Report Prepared for the World Bank Afghanistan Agriculture Sector Review, World Bank, Washington, DC.

Central Statistics Organization. 2016. "Statistical Indicators of Afghanistan." Government of Afghanistan.

Deichmann, U., F. Shilpi, and R. Vakis. 2008. "Spatial Specialization and Farm-Nonfarm Linkages." Policy Research Working Paper 4611, World Bank, Washington, DC.

FAOSTAT. 2016. Production Database. The Food and Agriculture Organization, Rome. Accessed from http://www.fao.org/faostat/en/#data

Foster, A.D., and M.R. Rosenzweig. 2004. "Agricultural Productivity Growth, Rural Economic Diversity, and Economic Reforms: India, 1970–2000." *Economic Development and Cultural Change* 52 (3): 509–42.

Gautam, M., and R. Faruqee. 2016. *Dynamics of Rural Growth in Bangladesh: Sustaining Poverty Reduction.* Washington, DC: World Bank.

Lanjouw, J., and P. Lanjouw. 2001. "Rural Non-Farm Employment: Issues and Evidence from Developing Countries." *Agricultural Economics* 26 (1): 1–24.

Mansfield, D., and A. Pain. 2007. "Developing Evidence-Based Policy: Understanding Changing Levels of Opium Poppy Cultivation in Afghanistan." AREU Briefing Paper (November), Afghanistan Research and Evaluation Unit, Kabul.

Pain, A. 2012. "Afghanistan's Opium Poppy Economy." Middle East Institute, Washington, DC. http://www.mei.edu/content/afghanistans-opium-poppy-economy on September 30, 2016.

Sen, B., M. Ahmed, M. Yunus, and Z. Ali. 2014. "Regional Inequality in Bangladesh: Re-Visiting the East-West Divide." BIDS-REF Discussion Paper, Bangladesh Institute of Development Studies, Dhaka.

Shilpi, F., and S. Emran. 2015. "Agricultural Productivity and Non-Farm Employment: Evidence from Bangladesh." Draft, Columbia University, New York.

SIGAR. 2014. "Quarterly Report to the United States Congress." Special Inspector General for Afghanistan Reconstruction. SIGAR, Washington, DC.

UNODC (United Nations Office on Drugs and Crime). 2015. "Afghanistan Opium Survey 2015." UNODC, Vienna.

U.S. Department of Agriculture. 2016. "Agricultural Total Factor Productivity Growth Indices for Individual Countries, 1961–2014," Economic Research Service, USDA, Washington, DC.

Ward, C., D. Mansfield, P. Oldham, and W. Byrd. 2008. "Afghanistan: Economic Incentives and Development Initiatives to Reduce Opium Production." World Bank, Washington, DC. https://openknowledge.worldbank.org/handle/10986/6272.

World Bank. 2016. "Fragility and Population Movement in Afghanistan." World Bank-UNHCR Policy Brief. World Bank, Washington, DC.

3 Employment, Skills, and Human Capital

IMPROVING VULNERABLE GROUPS' PROSPECTS FOR WELL-PAID WORK IN RURAL AFGHANISTAN

INTRODUCTION

Improving employment and livelihood opportunities for vulnerable groups such as youth, women, the landless, and the illiterate is key to developing successful poverty reduction programs. Because labor is often the only resource vulnerable groups have to offer, it is difficult for them to overcome poverty without sufficient employment opportunities. Programs and policies have supported vulnerable groups in Afghanistan with help from the World Bank and other donors under the Afghanistan Reconstruction Trust Fund initiative. Furthermore, the Afghanistan Rural Enterprise Development Program works to create sustainable nonfarm employment opportunities for small and micro entrepreneurs, and the National Solidarity Program III generated short-term employment for the poor and improved rural infrastructure, which helped over the longer term by improving overall connectivity and productivity. The On-Farm Water Management project is working to improve farmers' agricultural productivity and water use efficiency, and the National Horticulture and Livestock Project has been working to improve rural livelihood opportunities for smallholder farmers through horticulture and livestock.

To better design and implement programs and policies to improve livelihood opportunities for vulnerable groups, we analyzed the current sectoral and spatial distribution of their employment and livelihood opportunities. This chapter explores the link between education and employment in rural areas, focusing on the nature and structure of youth employment while discussing the challenge of jobs for young Afghans in rural areas. It then discusses developing better, more inclusive rural jobs for vulnerable groups. It uses education and employment data from the Afghanistan Living Condition Survey (ALCS) 2013–14 and data from secondary sources.

EDUCATION AND TRAINING FOR RURAL EMPLOYMENT: HUMAN CAPITAL DYNAMICS AND STRUCTURE FOR MORE AND BETTER JOBS IN RURAL AREAS

The human capital disadvantage faced by the rural poor cannot be over-emphasized. Although Afghanistan has dramatically increased investment in human capital since 2001, much remains to be achieved, especially among the poor. Children under 15 account for more than half of the poor population. About 76 percent of the poor older than 15 are illiterate (compared with 63 percent of the non-poor), and only 7 percent have completed primary education (World Bank 2015, 6). In this backdrop, understanding the dynamics of youth and literate workers' employment across regions is crucial to identifying the current state of better jobs in rural Afghanistan. Educated workers generally prefer to engage in nonfarm activities, which tend to be more economically rewarding than farm activities. The lack of opportunities in the nonfarm sector, however, may lead skilled workers in rural areas to engage in less economically rewarding agriculture production activities, which can discourage prospective workers from acquiring new skills. The youth workforce is generally more literate than their elder cohorts (figure 3.1), and are more interested in better-paid jobs in the nonfarm sector that require more skills. On average, male workers are more educated than female workers—a difference also true of the young generation of workers.

Map 3.1 plots the average years of schooling for male and female workers in rural areas to understand the spatial patterns of schooling and human capital across the country. (Here, "workers" are people aged 15–65 who have participated in the labor force.) The maps reveal that education is not equally spread across provinces. Schooling is high for male and female workers in the provinces in and around Kabul, but is lower in the provinces in the south and west central regions. Chapter 2 noted that shares of nonagricultural employment and income are generally high in provinces around the capital, meaning there is a positive association between worker's education level and their nonfarm engagement.

FIGURE 3.1

Average years of schooling of rural workforce, 2011–12 and 2013–14

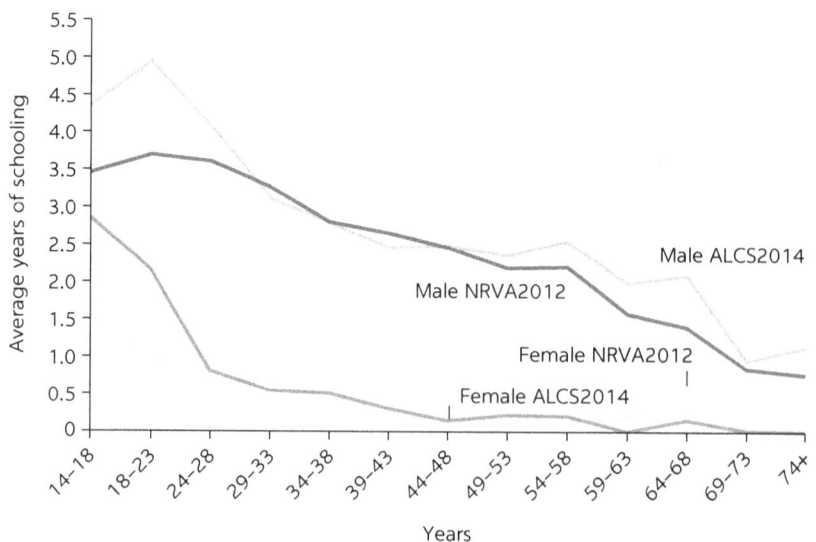

Sources: Based on data from NRVA 2011–12 and ALCS 2013–14.

MAP 3.1

Spatial patterns of schooling of male and female workers, 2011–12 and 2013–14

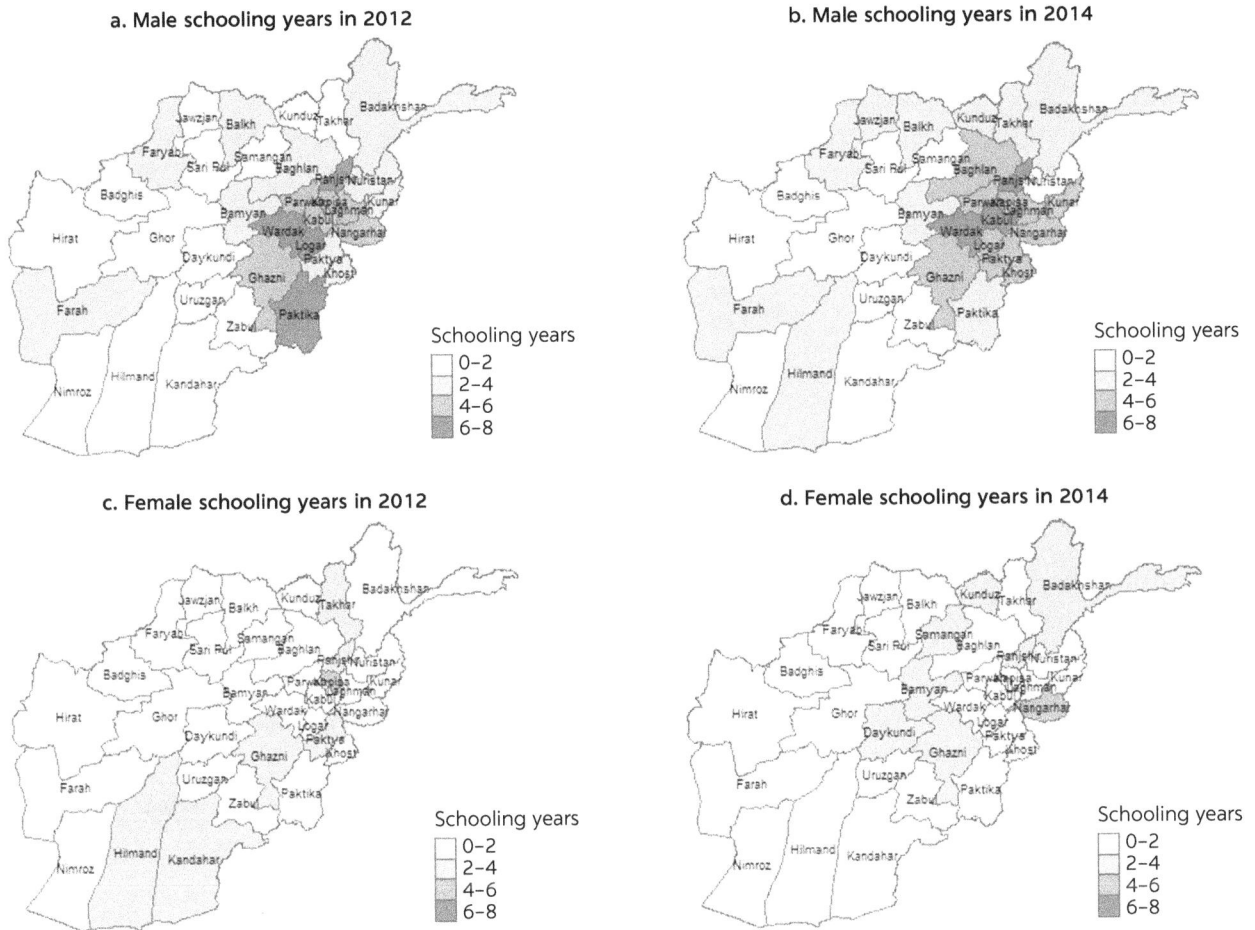

a. Male schooling years in 2012

b. Male schooling years in 2014

c. Female schooling years in 2012

d. Female schooling years in 2014

Sources: Based on NRVA 2011–12 and ALCS 2013–14.

Education and Employment in Rural Areas

There are substantial differences between the labor force participation rate (LFPR) of literate and illiterate workers (figure 3.2). In general, literate people participate more. The ALCS 2013–14 revealed that the LFPR for literate and illiterate people was about 66 percent and 50 percent, respectively. The primary reason for this striking difference is low female literacy and women's low participation in the labor force. The low LFPR among illiterate people may also reflect the low LFPR among women.

The LFPR for males is much higher than for females, regardless of literacy, and the LFPR for illiterate males is consistently higher than for literate males (table 3.1). These patterns are consistent across all regions. Reduced employment opportunities in the formal sector because of economic decline may discourage literate males from joining the labor force.

Notably, male and female LFPR patterns by literacy status are completely different. The LFPR among literate females is consistently higher than among illiterate females. ALCS 2013–14 data show that the LFPR was 34 percent for literate women and about 28 percent for illiterate women (table 3.1). Literate women's LFPR is higher in most regions. In 2014, the LFPR was higher among illiterate women only in the west central and south regions. Although the lack

FIGURE 3.2

Labor force participation rate in 2013–14: literate versus illiterate workers

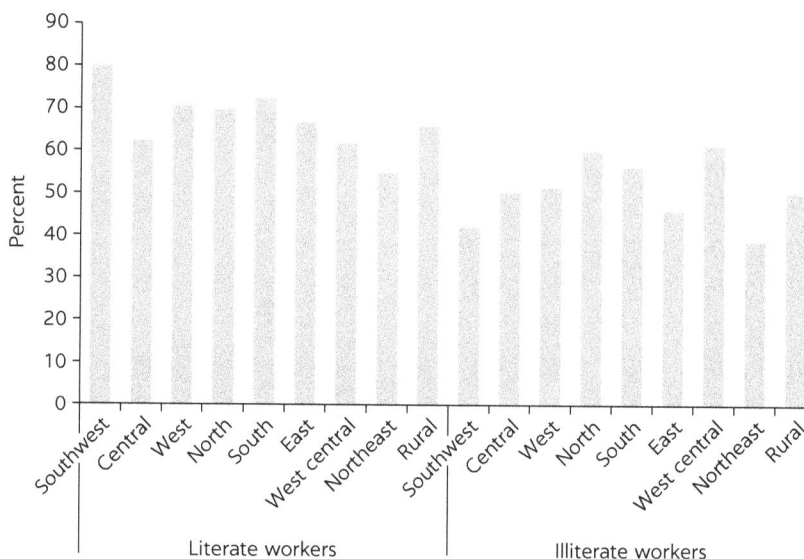

Source: Based on ALCS 2013–14.

TABLE 3.1 **Male and female labor force participation rate by literacy, 2013–14 (percent)**

REGION	MALE		FEMALE	
	Literate	Illiterate	Literate	Illiterate
Southwest	85.5	85.3	23.2	6.0
Central	71.0	83.5	34.0	34.7
West	82.8	84.6	40.4	25.4
North	77.1	86.0	49.3	41.3
South	81.2	84.4	25.9	44.1
East	73.3	87.6	31.0	23.3
West central	70.3	85.7	39.4	43.5
Northeast	66.6	84.2	24.8	5.7
Rural	75.1	85.1	34.3	27.7

Source: Based on ALCS 2013–14.

of sufficient and suitable job opportunities may have discouraged rural women from participating more in the labor force in 2014, the stark differences in LFPR have important policy implications. Literacy improved the LFPR of the female working-age population; therefore, the government and policymakers should increase efforts to improve the human capital of Afghanistan's female population.

Despite their low LFPR, illiterate people experience unemployment more severely than literate people (figure 3.3). This is true for all regions, though it is more prominent in the west central, west, central, north, and northeast regions. The underemployment situation is also more acute among illiterate workers in most regions. The agriculture and livestock sectors account for

FIGURE 3.3

Unemployment and underemployment rates in 2013–14: literate versus illiterate workers

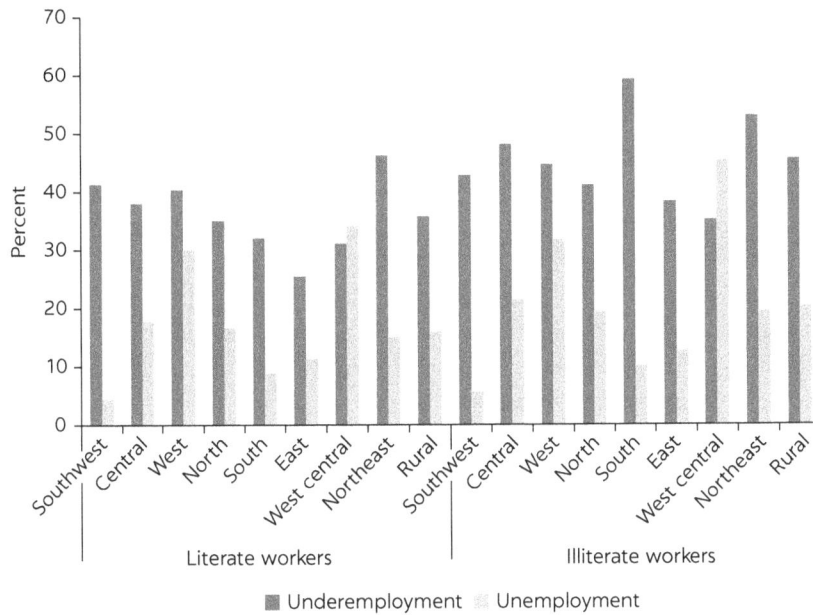

Source: Based on ALCS 2013–14.

FIGURE 3.4

Sector of employment in 2013–14: literate versus illiterate workers

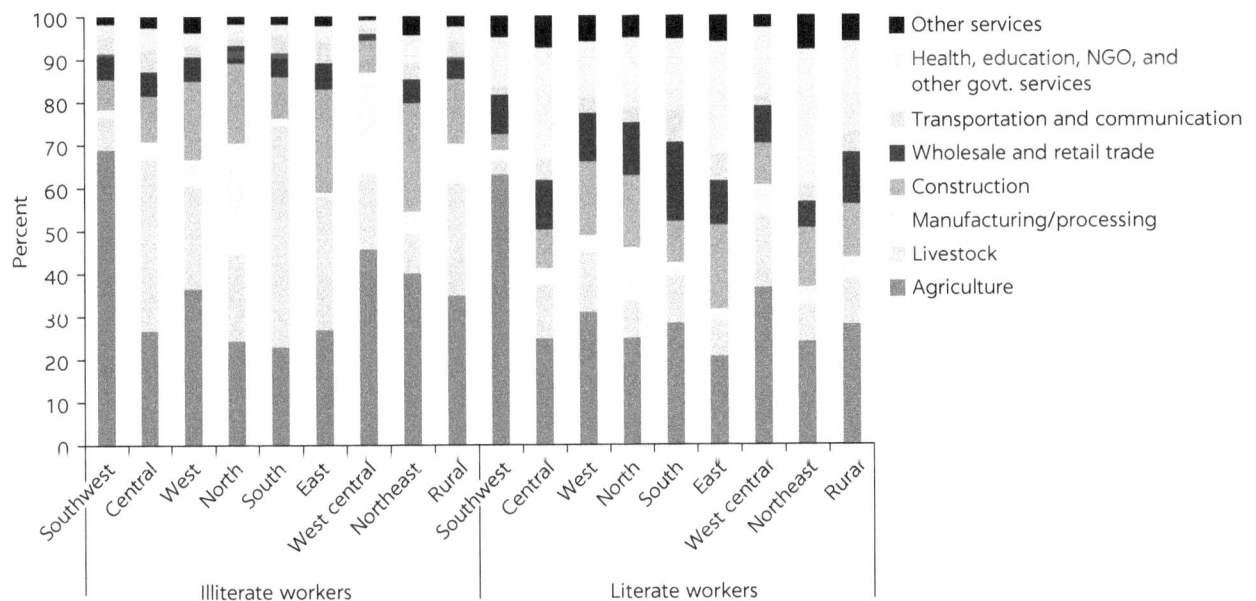

Source: Based on ALCS 2013–14.
Note: NGO = nongovernmental organization.

61 percent of employed illiterate people, but only 38 percent of literate people (figure 3.4). Thus, the informal nature of the jobs done by the illiterate workforce and the recent decline in agricultural performance may be the key factors behind the high unemployment and underemployment among illiterate workers in rural areas.

There are distinctive features in the sectoral distribution of literate and illiterate workers in rural areas. Expectedly, literate workers are more involved in nonfarm activities and illiterate workers are more involved in on-farm activities (figure 3.4). The major employment sectors for illiterate workers are agriculture (34.7 percent), livestock (26.5 percent), construction (15.1 percent), and manufacturing (8.8 percent); for literate workers, they are agriculture (28.1 percent), service (21.1 percent), construction (12.4 percent), wholesale and retail trade (12 percent), and livestock (10.5 percent). There are significant regional variations in the sectoral distribution of employment for literate and illiterate workers. For example, about two-thirds of the illiterate workforce in the southwest is employed in the agricultural sector. In the south, 52 percent of the illiterate workforce is employed in the livestock sector. In the west central, northeast, and west regions, agriculture is the dominant source of employment for illiterate workers. Livestock is the dominant sector in the central and east regions.

Manufacturing and processing employs many illiterate workers in the north and west central regions. The construction sector generates a relatively higher number of jobs for the illiterate workforce than for the literate workforce; for both groups, employment in the sector is clustered in the northeast, east, north, and west. The illiterate workforce's share of employment in the formal sector is very low in each region, while it is relatively high for the literate workforce in many regions. For example, about one-third of the literate labor force in the northeast is employed in the formal sector. In the central and east regions, where Kabul is situated, about one-quarter of literate workers are employed in the formal sector.

The informal nature of the illiterate workforce's jobs is also reflected in the type of employment (figure 3.5). While about 27 percent of literate workers were employed in the formal sector in 2014, illiterate workers accounted for only about 5 percent. Informal work makes illiterate workers more vulnerable,

FIGURE 3.5

Type of employment in 2013–14: literate versus illiterate workers

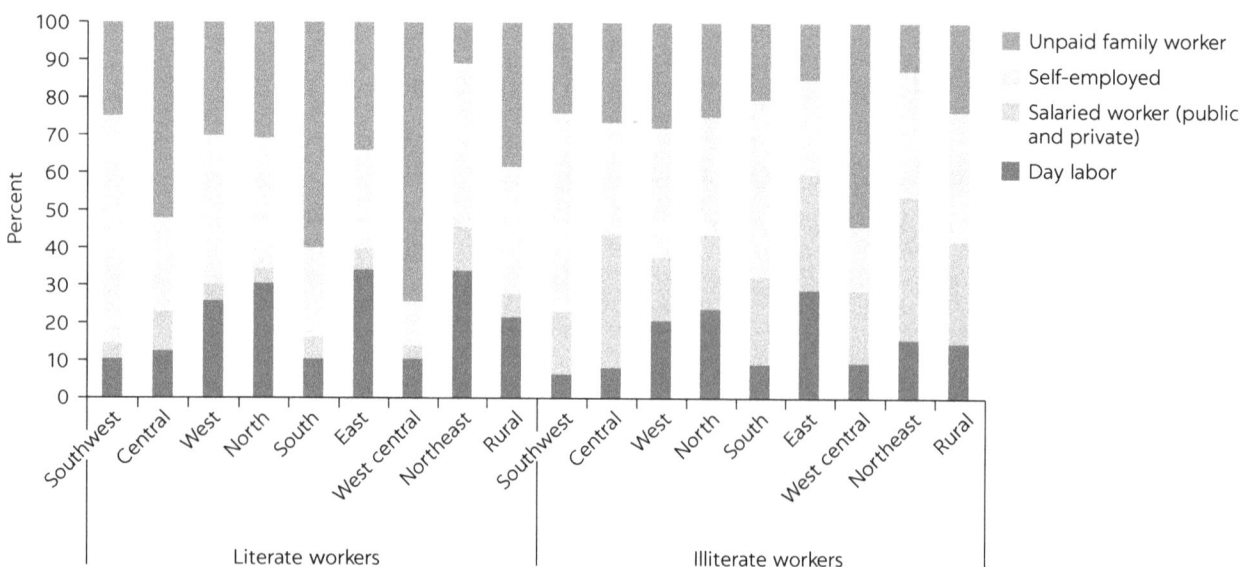

Literate workers — Illiterate workers

Legend: Unpaid family worker; Self-employed; Salaried worker (public and private); Day labor

Source: Based on ALCS 2013–14.

especially to weather and other natural shocks. For example, in 2014, about 78 percent of rural households with a literate head of household reported that they had experienced at least one shock (of 30 shocks listed in the questionnaire) in the past year, but for households with an illiterate household head, the figure was 84 percent. In 2014, most illiterate workers were unpaid family workers, followed by the self-employed and day laborers; most literate workers were self-employed, followed by salaried workers and unpaid family workers. Also in 2014, 39 percent of the illiterate workforce was involved in economic activities as unpaid family workers; the figure was 24 percent for literate workers. Therefore, literacy and skills development are important factors in the reduction of vulnerability through paid jobs.

YOUTH WORKERS: THEIR EMPLOYMENT DYNAMICS IN RURAL AFGHANISTAN

Interest in job creation has never been higher—and rightly so, as it is estimated that 660,000 new workers will join the labor force every year (see chapter 2). The expansion of opportunities to create more, sustainable, and inclusive jobs for the youth remains key. However, the security situation remains fluid, and the conflict's intensity has escalated in recent years. This volatile situation often poses the toughest challenges to the implementation of interventions that can support the youth in remote areas. As a result, the economy's capacity to productively employ the youth is central to the prospects for sustainable development and poverty reduction, as well as to the emergence of a peaceful and stable Afghanistan.

Revised projections of gross domestic product growth (estimated at 3 percent in 2017 [Asian Development Bank 2016]), driven partly by expectations of continuing conflict in several parts of the country, further magnify the challenge of youth unemployment. Slowing growth limits the economy's ability to absorb and reward skilled youth. Therefore, the potential for job creation for young people lies in two areas: sectors that are likely to grow even in a slowing economy and self-employment in parts of the nonformal sector that are likely to have local demand.

Young workers—those under 25—often have different employment preferences than adult workers. In most regions, they have a lower LFPR (figure 3.6). This could be because some young people are in school, pursing higher education. Spatial variations exist: Compared with adults, youth have a much lower LFPR in the northeast and central regions, but it is higher in the north, west, and southwest.

Despite their low LFPR, youth are often unable to find paid employment and have a much higher unemployment rate than adult workers. For example, the 2014 unemployment rates for youth and adults, respectively, were about 24 percent and 16 percent (figure 3.7). In some regions, youth unemployment rates are markedly high. About half of the youth workforce in the west central region is unemployed, and the rate surpasses 25 percent in the west, northeast, and central regions. This high unemployment may be due to lack of sufficient new and more job opportunities and skills gaps.

The underemployment situation is better for employed youth than for their adult counterparts (figure 3.7). The underemployment rate among adult workers is particularly severe in the northeast region, where about 56 percent are

FIGURE 3.6

Labor force participation rate in 2013–14: adult versus youth workers

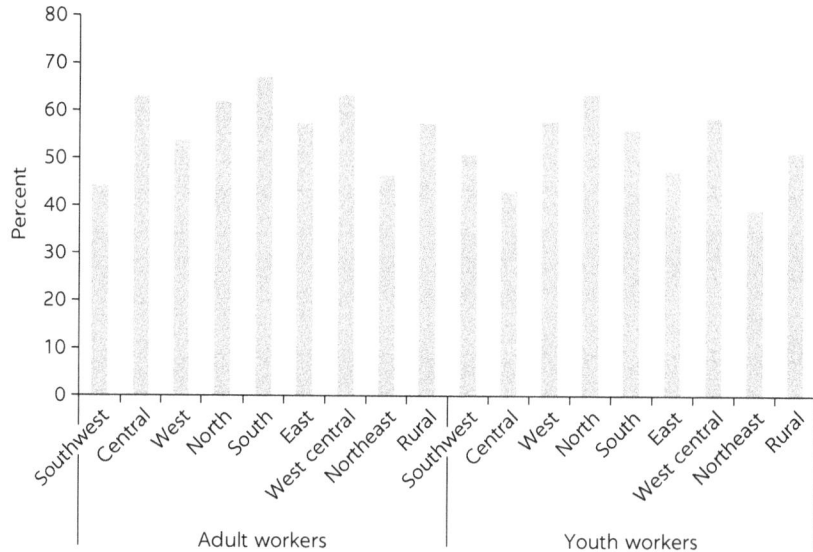

Source: Based on ALCS 2013–14.

FIGURE 3.7

Unemployment and underemployment rates in 2013–14: adult versus youth workers

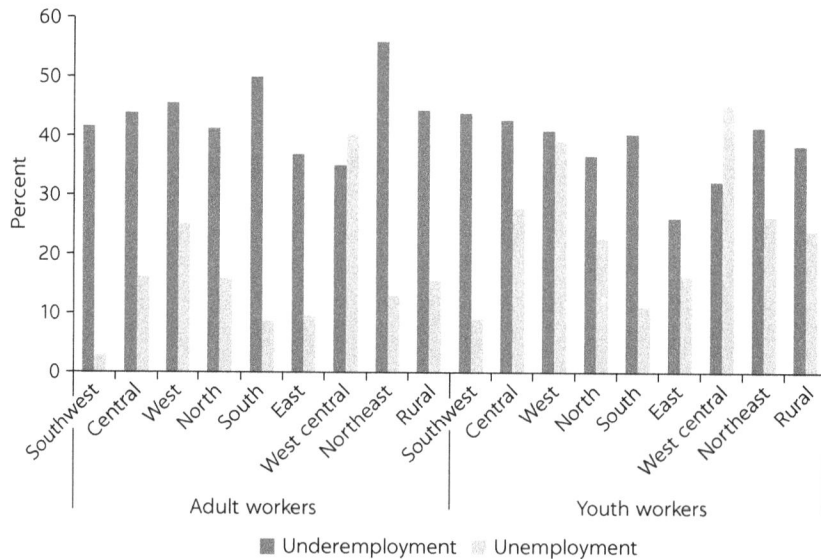

■ Underemployment ▨ Unemployment

Source: Based on ALCS 2013–14.

underemployed. Spatial patterns exist: The rates for both groups are relatively low in the west central region, but youth underemployment is also relatively low in the east.

Our analysis found that the youth workforce is more literate than the adult workforce, and that youth workers are therefore expected to participate more in the nonfarm activities of the rural economy (see chapter 2).

Nevertheless, the sectoral distribution of youth and adult employment is not much different. For example, in 2014, about one-third of both youth and adult workers were involved in agriculture. However, youth were employed more in livestock activities (24 percent vs. 20 percent) and in manufacturing/food processing (11 percent vs. 6 percent). However, youth's share of formal sector employment (e.g., health, education, NGOs, and other public services) was lower (8 percent vs. 10 percent). Youth are also working less in the transportation sector.

There are significant variations in the sectoral distribution of youth workers across regions. While about two-thirds of youth employment in the southwest is in agriculture, the sector accounts for about one-quarter of employment in the central, north, south, and east regions. Livestock accounts for a sizable share (20–36 percent) of youth employment in these regions. Although the manufacturing and processing sector accounts for 11 percent of total rural employment, its share of youth employment is more than 25 percent higher in the north and west central regions (figure 3.8). The types of employment also vary among adults and youth. Respectively, the dominant employment type for adult and youth workers is self-employment and unpaid family workers (figure 3.9).

In 2014, the share of self-employment and unpaid family work was about 66 percent among adult workers and about 71 percent for youth workers. The labor status for both groups varies from region to region. In the southwest and the west central regions, more than 80 percent of employed youth is self-employed or an unpaid family worker. The share is about 55 percent in the east and northeast regions.

FIGURE 3.8

Sector of employment in 2013–14: adult versus youth workers

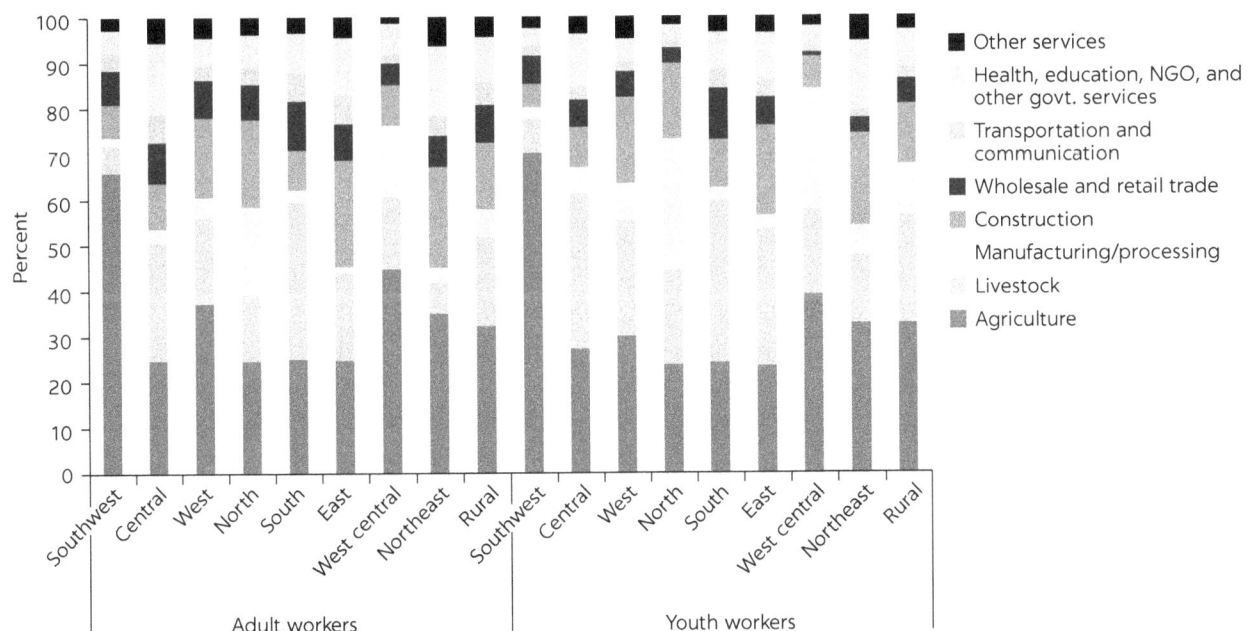

Source: Based on ALCS 2013–14.
Note: NGO = nongovernmental organization.

FIGURE 3.9

Type of employment in 2013–14: adult versus youth workers

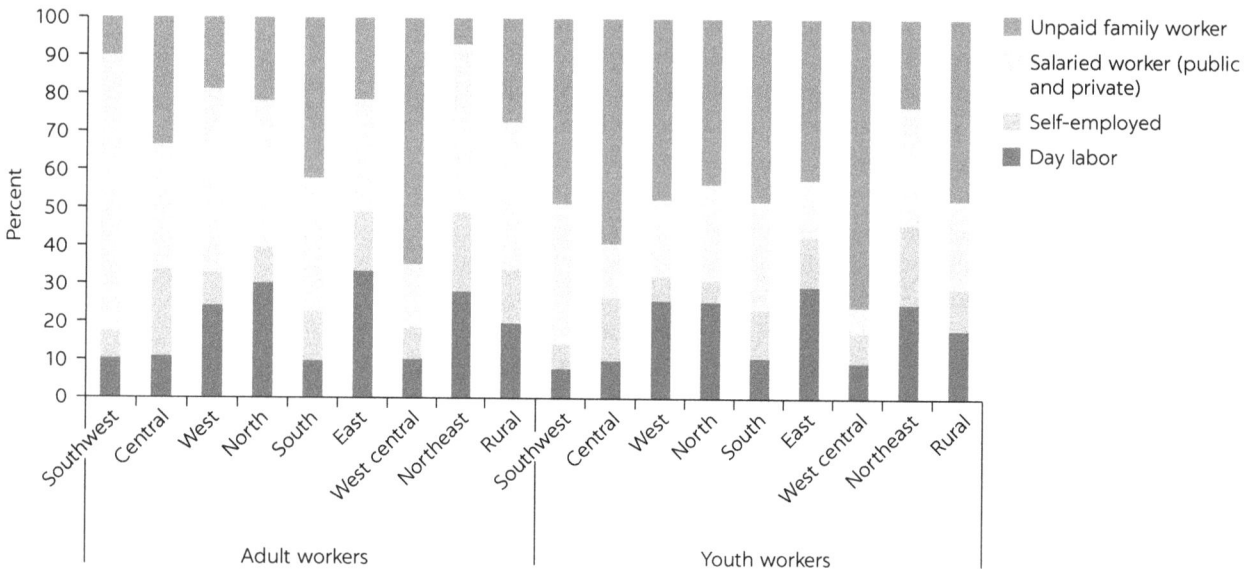

Source: Based on ALCS 2013–14.

INCLUSIVE JOBS IN RURAL AFGHANISTAN: EMPLOYMENT DYNAMICS OF THE LANDLESS, WOMEN, AND THE BOTTOM 40 PERCENT OF INCOME EARNERS

To achieve poverty reduction and inclusive economic growth, a longer-term job strategy must be part of Afghanistan's policies and actions for the creation of sustainable employment in rural areas. It is particularly important to understand the employment dynamics of the most vulnerable groups. Although livelihood vulnerability is a complex issue, this analysis will focus only on landlessness, gender, internally displaced people and returnees, and the bottom 40 percent of income earners.

Landless households: creating opportunities for a group with limited options

In 2012, about 32 percent of rural households were landless. By 2014, this had increased to 37 percent. Landlessness is especially acute in remote border provinces (map 3.2), and it is uncommon for these households to have access to agricultural land (i.e., land to lease) (map 3.3).

Without access to agricultural land, workers from landless households can engage in the agriculture sector as wage laborers, or in the nonfarm sector as wage laborers, skilled workers, or self-employed entrepreneurs. As noted in chapter 2, the share of nonagricultural employment is generally low in the remote provinces, and landless workers there are more likely to be employed as agricultural wage laborers. The employment dynamics and the income dynamics of landless rural households offer insights into their income-generating activities and employment preferences.

The LFPR among working-age adults from landless households is slightly lower than for those from landowner households (figure 3.10). About 56 percent of working-age adults from landowner households join the labor force;

MAP 3.2
Percent of landless households in rural Afghanistan

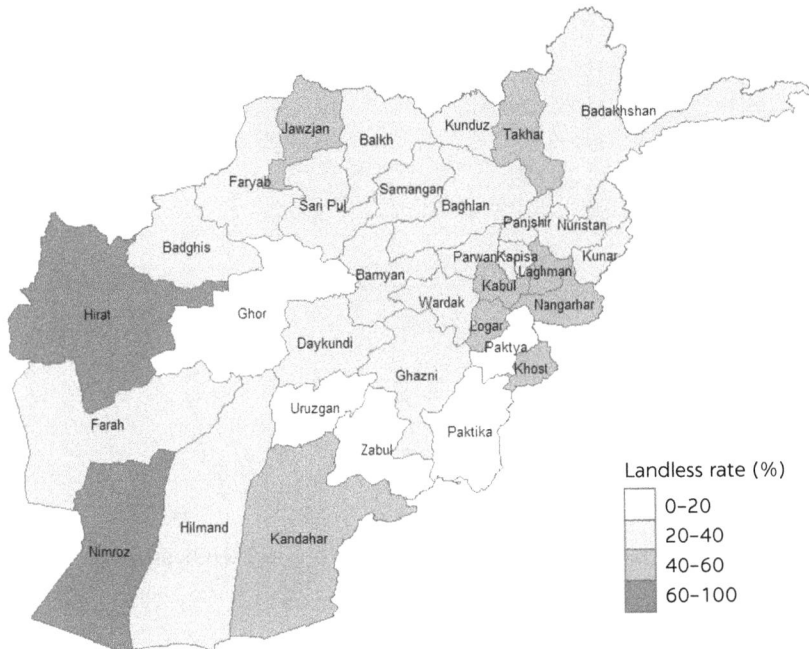

Source: Based on ALCS 2013–14.

MAP 3.3
Percent of landless rural households with land access

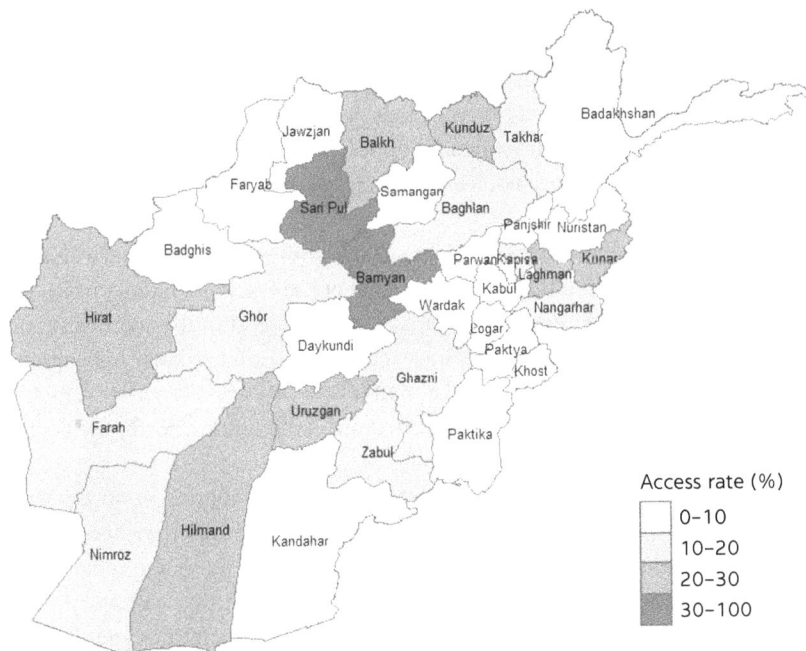

Source: Based on ALCS 2013–14.

FIGURE 3.10

Labor force participation rate in 2013–14: landless versus landowner

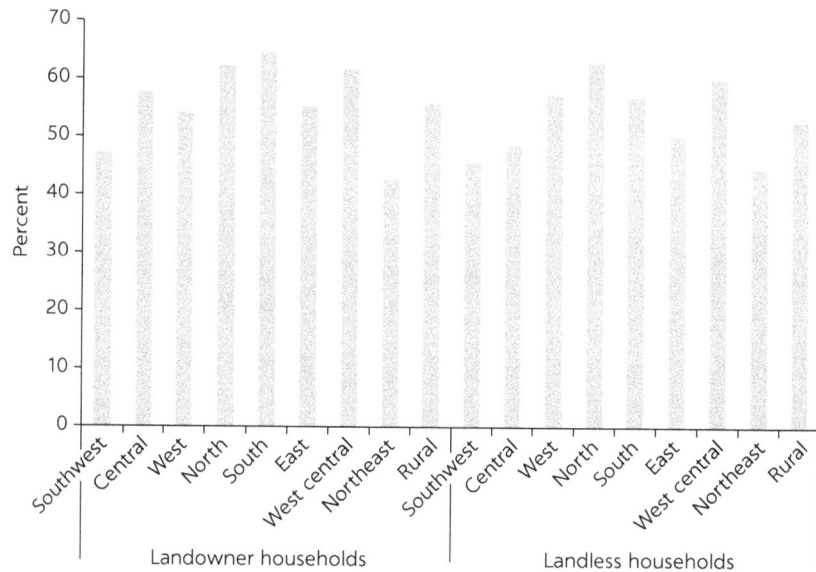

Source: Based on ALCS 2013–14.

it is 53 percent for landless households. Spatial variation also exists in the LFPR of adults from landless households: More than 60 percent join the labor force in the north and west central regions, but the LFPR is less than 50 percent in the southwest, central, and northeast regions. Their LFPR is lower than for adults from landowner households in the central, south, and east regions. Their LFPR is low in the central region (48.6 percent), where Kabul is located, even compared with adults from landless households in other regions. Although the literature demonstrates that proximity to urban areas generally does matter for nonfarm employment opportunities in rural areas, it appears that proximity to Kabul has not led to an increase in the LFPR among workers from landless households.

The unemployment rates of workers from landless and landowner households do not differ much in rural areas. There are some spatial variations, however. For example, the rate for both groups is lowest in the southwest and highest in the west central region. However, the rates are markedly high among landless workers in the west central region (about 50 percent) and west region (more than one-third) (figure 3.11). Still, the rate in the central, north, and northeast regions is lower for landless workers.

Figure 3.11 also shows that the underemployment situation is more severe for workers from landowner households. About 50 percent are underemployed, compared with about 33 percent for workers from landless households. The difference is high in the southwest, central, west, north, and south regions. The stark differences mainly originate from the type of work that the groups pursue. The situation is generally less severe among day laborers and employees in the formal sector, and more severe among the self-employed and the voluntary family workers who are primarily employed in agriculture and livestock. More than half of the landless workers are day

FIGURE 3.11

Unemployment and underemployment rates in 2013–14: landless versus landowner

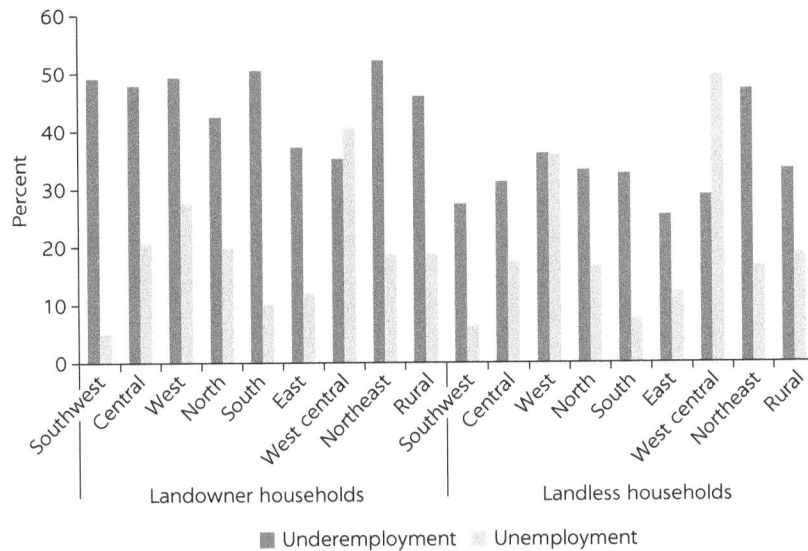

Source: Based on ALCS 2013–14.

laborers or salaried workers, compared with about one-quarter of workers from landowner households. Workers from landowner households are more likely to engage in agricultural activities as self-employed farmers, and are thus more likely to be underemployed.

As expected, the sectoral distribution of employment varies substantially between the groups. Workers from landowner households are more involved in the agriculture and livestock sectors, while those from landless households are more engaged in nonfarm activities. Construction is the most important employment sector for landless workers (24 percent), followed by agriculture (17.8 percent) and livestock (14.2 percent). In 2014, about 12 percent of landless workers were involved in the health, education, NGO, and government services sectors, compared to about 8 percent of landowners. About 62 percent of landowner workers were employed in the agriculture and livestock sectors, compared with about 32 percent of landless workers.

The sectoral distribution of employment of both groups varies significantly across regions. About 35.7 percent of workers from landless households are employed in agriculture in the southwest; in the south, it is only about 8.3 percent (figure 3.12). Livestock is the dominant employment sector for landless workers in the south. Construction is an important sector for these workers in most regions, except in the west central region. Manufacturing and processing is an important sector for both groups in the north and west central regions. Employment status also varies significantly: About three-quarters of landowner workers are self-employed or unpaid family workers, compared with less than half of landless workers (figure 3.13). Most landless workers are employed as day laborers or salaried workers. (About one-third are day laborers, compared with about 13 percent of landowners.)

FIGURE 3.12

Sector of employment in 2013–14: landless versus landowner

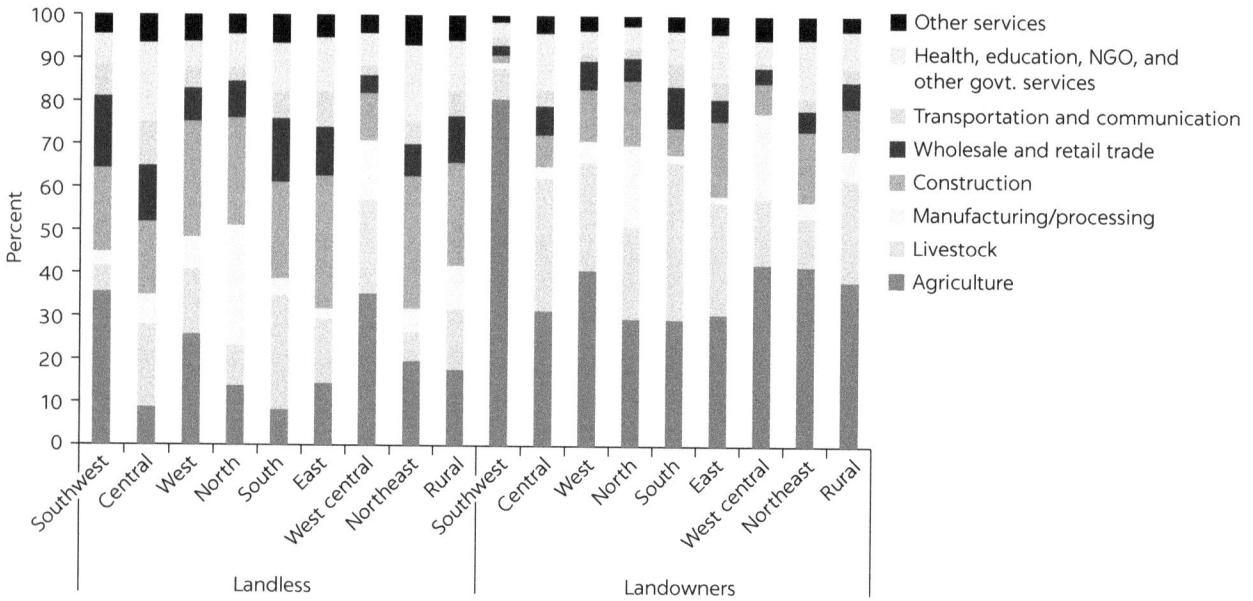

Source: Based on ALCS 2013–14.
Note: NGO = nongovernmental organization.

FIGURE 3.13

Type of employment in 2013–14: landless versus landowner

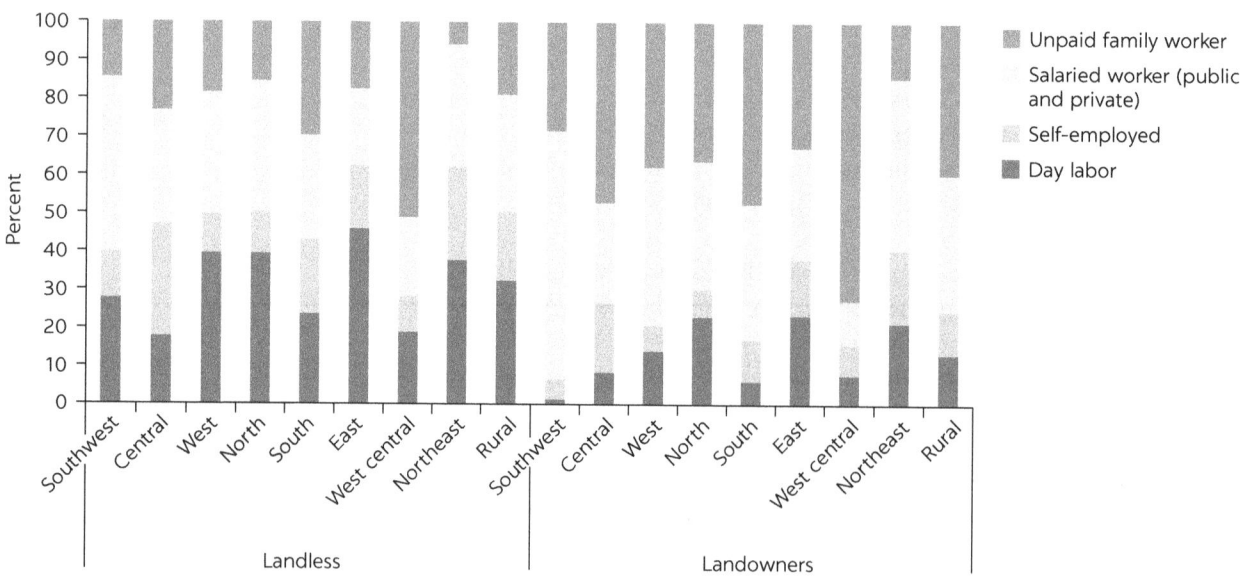

Source: Based on ALCS 2013–14.

BOX 3.1

Spotlight: Land in Afghanistan

The current land situation. Afghanistan is predominantly a mountainous, dry country with limited fertile land. Five percent of the land area is under irrigation, and another 7 percent is available for rain-fed farming every two, three, or more years (FAO 1999). Forty-five percent of the land is agro-ecologically classified as rangeland. Another 37 percent is categorized as "barren," but is usable as pasture on a seasonal basis (Wily 2004a). Among the estimated 32 million rural dwellers, there are 2.4 million Kuchis, mainly Pashtun descent, 60 percent of whom remain fully nomadic livestock keepers. Most live in the south and east regions; over the past century, they have been moving more and more deeply into the central west and north for summer pasturing—a source of contention with local populations (Wily 2004b).

Land reforms: legal framework. There are more than 30 laws, decrees, and documents related to formal land administration in Afghanistan.

Administrative-based land registration and titling. Afghanistan is one of the very few countries where land registration falls under the judiciary. The Council of Ministers recently approved transferring the land registration and titling mandate to an administrative system, and the Afghanistan Independent Land Authority (ARAZI) is planning to develop a five-year strategic plan by mid-2017 with a 50-year roadmap on land registration. The administrative land registration and titling model will be piloted in Kabul and Herat, based on a model adapted from Turkey. It is expected that the new system will significantly cut time, costs, and red tape.

Status of the land sector:

- Total area. 652 000 sq. km, of which 12% is arable.
- 3% is forested, 46% is permanent pasture, and 39% is mountainous and not usable for agriculture.
- About 70% of rural households have access to land (through ownership, lease, and/or sharecropping).
- The government claims ownership of more than 80% of land resources.

- Around 44% of households have access to irrigated land (averaging 1.2 ha); 26% have access to rain-fed land (averaging 3.2 ha).
- The court (judicial) registration system: Ownership rights are clarified in the courts. Deeds are registered by the courts based on a judicial process. A small proportion of landowners hold court-registered deeds.
- The courts are the final authority for land conflict resolution, but with the general lack of legal documents, the system tends to favor the powerful.
- Legal reforms initiated: The percent of households with access to land increased between the NRVAs in 2005, 2007–08, and 2011–12, but average access to irrigated land is decreasing.
- Around 34% of the land was covered by cadaster in the 1960s and 1970s; few hold formal land titles, land registration has not been updated for several generations, and there are many fake and overlapping deeds.
- Registering property: The 2017 Doing Business report indicates that, on average, the process in Kabul involves nine procedures, 250 days, and costs 5% of the property value.

Land conflict resolution. Land disputes have long driven violent conflict in Afghanistan. Widespread poverty and a scarcity of productive land generates intense competition among communities, ethnicities, and tribes for land and resources. Disputes over access to land and water are a major source of inter- and intra-communal conflict, and can have violent ramifications (Wily 2013, 3–4). These dynamics are exacerbated by historically unequal land distribution and periodic forced redistribution and resettlement of groups from particular ethnicities for political control (see box 3.4).[a]

Land Policy and Regulatory Framework: ARAZI

- In 2013, ARAZI was transformed into an independent public institution with the idea of

continued

Box 3.1, *continued*

developing it into a "one-stop-shop" for all land administration matters.

- A Land Dispute Resolution Department was created, and procedures, policies, manuals and regulations were drafted.
- ARAZI significantly pushed for amendments to land management and land acquisition laws.
- ARAZI's goal: "To be Afghanistan's prime and sole, independent land administration and management body, managing state owned lands, and providing land related services and information to citizens, institutions and investors."

Potential Interventions

- Strengthen ARAZI to become and remain a well-managed and "clean" institution
- Shift from court-based registration to an administrative system and implement work in prioritized manner
- Formalize out-of-court dispute settlements through a recognized process

Legal and policy reform. The law is so narrowly written that, in most cases, the state is a threat rather than a protector of rights. Because customary ownership and long-standing communal ownership or usage rights are not recognized in Afghan land management law, fundamental conflicts between the state and landowners and tribes are inevitable. These two issues doom any initiative to increase registration and titling. Reforming the land law to enable legal recognition of communal land rights would significantly expand the state's ability to productively engage with some of the most common sources of land disputes.

In addition to these legal reforms, a full-scale cadastral survey and a comprehensive national land titling and registration program are essential. A short-term priority to enable these goals would be to build the internal capacity of all state actors engaged in land management. A vital—but time-consuming aspect—of the administrative land registration and titling model pilot is developing ARAZI's internal

capacity. Such efforts must continue and be addressed at the national, district, and provincial levels.

Last, an important guiding principle for both short- and long-term goals should be greater receptivity to community interests and land management solutions that respond to the reality of land tenure. In the pilot, community-based dispute resolution processes were allowed to feed into land registration and identification processes. This was useful, but the validity of such processes is still controversial under Afghan law. Working with community elders or *shuras* on a systematic level would require greater legal and policy development and reform. Land management authorities must identify a way to involve community preferences in any formalization process. Without that, the problems that arise from the current unregulated system are likely to continue.

Resolving land disputes involving refugee/ internally displaced people. The Karzai administration issued four land decrees, two of which related to land disputes arising during the absence of owners since April 27, 1978 (e.g., with refugees and internally displaced people). The first established a single Property Disputes Resolution Court in Kabul in 2002, which has since been replaced with a two-tier system. The second provided two courts, one to deal with disputes in Kabul province and one for outside Kabul.[b] If the government is one of the disputants, these courts may not hear the case. This is problematic in rural areas where the government's claim to lands (variously defined as public or government land) is central to the problem.

Addressing Land Conflict

- Almost 33% of private and state lands have been surveyed.
- State mechanisms are even less able to sustainably settle disputes given their limited presence, poor enforcement capability, bad reputation (due to corruption and land grabbing), and the widespread lack of authentic title deeds.
- Land disputes are a primary driver of conflict.

continued

- Historically, land disputes were mediated through community-based dispute resolution, but the two decades of conflict and instability has weakened community social structures. Socioeconomic changes and the ongoing insurgency and displacement since 2002 have further destabilized traditional mechanisms.
- The Land Management Law creates a catch-22: To establish formal legal ownership based on customary documents, one must already have formal legal ownership established in an original formal land document.

Recommendations

- National cadastral mapping and land surveys are needed to clarify land ownership and user interests.
- The ARAZI Land Dispute Resolution Department should continue its efforts to develop cooperation with community actors (community-based dispute resolution).
- ARAZI should be empowered to conduct land registration on a more widespread basis.
- Finalize/amend land laws/policies, regulations, and guidelines.

Source: World Bank 2017.
a. Afghanistan Council of the Asia Society 1978.
b. Decree 136 (19/6/1381) 2002 in Gazette 804, now replaced with Decree 89 (9/9/1382) 2003 Regarding the Creation of a Special Property Disputes Resolution Court.

Gender and employment structure in rural areas: enabling women with increased access to income-generating opportunities

Although women have always played a key role in all dimensions of agricultural production, most of their labor has been unpaid. In 2005, the World Bank's Gender Country Assessment (World Bank 2005) looked at the role of women in Afghanistan's future, focusing on three points: the mostly agrarian nature of the economy, women's largely unpaid participation in horticultural and livestock employment, and the gender division of labor in the agricultural sector. The assessment stated that most rural households' livelihoods were a mosaic of jobs, such as crop production, livestock, wood-cutting, labor, and other small-scale activities. It also indicated that women frequently contributed with economic activities beyond agriculture. The World Bank's latest Gender Country Assessment (World Bank 2014b) took stock of achievements and ongoing challenges in the role of women in the country's future, and concluded that the government of Afghanistan had made important commitments to women. While these have translated into demonstrable progress in some sectors, such as health and education, they have been less visible elsewhere, such as work and employment. In agriculture, though women continue to play an important role in production processes and household farming strategies, their work is still often unpaid and devoted to activities not related to decision-making or brokering trade exchanges with the market. Therefore, the sector's gender structure and women's role in jobs remains mostly unchanged.

To identify pathways for creating inclusive and sustainable jobs for Afghan women, especially the poorest, we look at the current employment situation of female workers in rural area. The employment scenarios of the female workforce in rural areas give insights into their employment preferences and potential policy options for inclusive job creation. Since the fall of Taliban regime in 2001, Afghan women have started to be involved in livelihood and income-generating activities out of their households. Yet, after 16 years, the share of women in the rural labor force is the lowest in South Asia; women face social and religious barriers to working outside the home, as well as labor market constraints (e.g., a lack of sufficient work opportunities).

Female LFPR varies significantly. For example, in nine provinces, less than 10 percent of rural households have female workers (map 3.4). However, nine provinces have female working members in more than 40 percent of households. Women's LFPR is very low in southwest provinces (Helmand, Kandahar, Zabul) and northeast provinces (Baghlan, Takhar, Kunduz) (map 3.3 and map 3A.1). In northwest provinces (Daykundi, Herat, and Badghis), less than 20 percent of households have female workers. While provinces around Kabul have a higher proportion of households with female workers, in Kabul province itself, the figure is 20–30 percent. Faryab, Jawzjan, Khost, Kunar, Nuristhan, Paktika, Paktya, and Wardak provinces have a high percentage of households with female workers. The female LFPR in rural areas is quite lower than the male LFPR, especially in the southwest and northeast regions (figure 3.14); however, it is relatively high in the north, south, west central, and central regions.

Despite rural women's lower LFPR, their unemployment rate is more than double that of males (figure 3.15). For example, in 2013–14, the rate was 15 percent for males and about 29 percent for females. The rates are acute for both men and women in the west, west central, and north regions. The female unemployment rate is more than 50 percent in the west and west central regions; it is also high in the northeast and central regions. The male unemployment

MAP 3.4

Percent of rural households with female workers, 2013–14

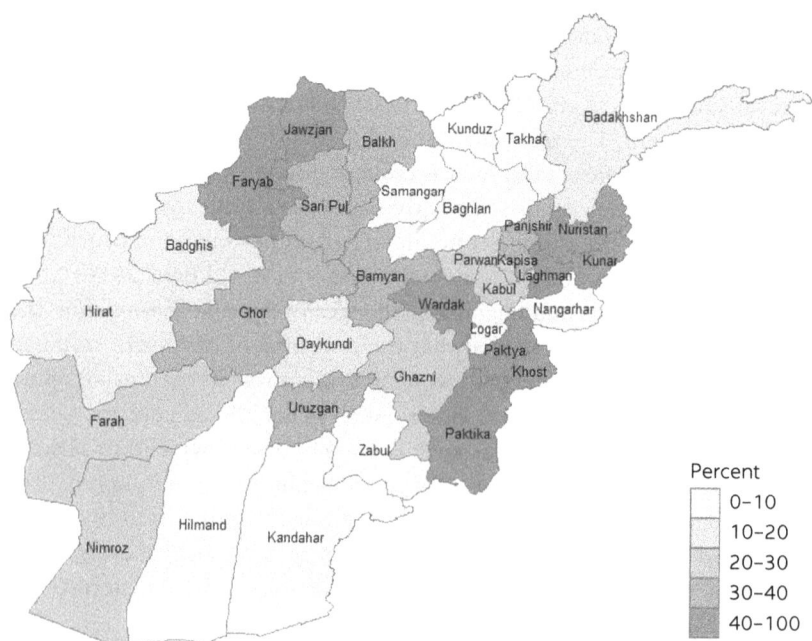

Source: Based on ALCS 2013–14.

FIGURE 3.14

Male and female labor force participation rate, 2013–14

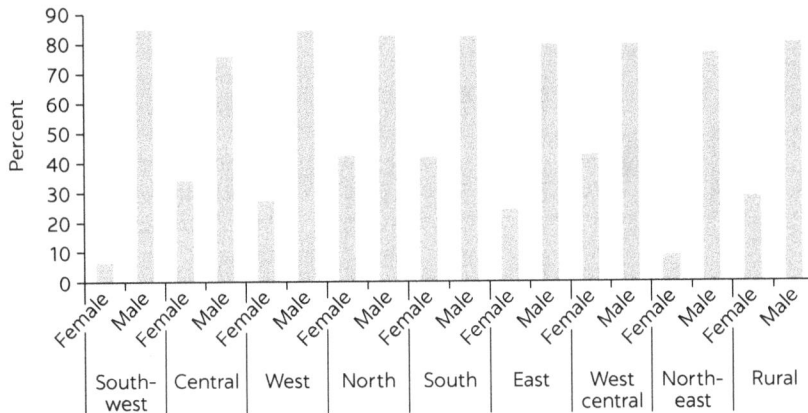

Source: Based on ALCS 2013–14.

FIGURE 3.15

Male and female unemployment and underemployment patterns, 2013–14

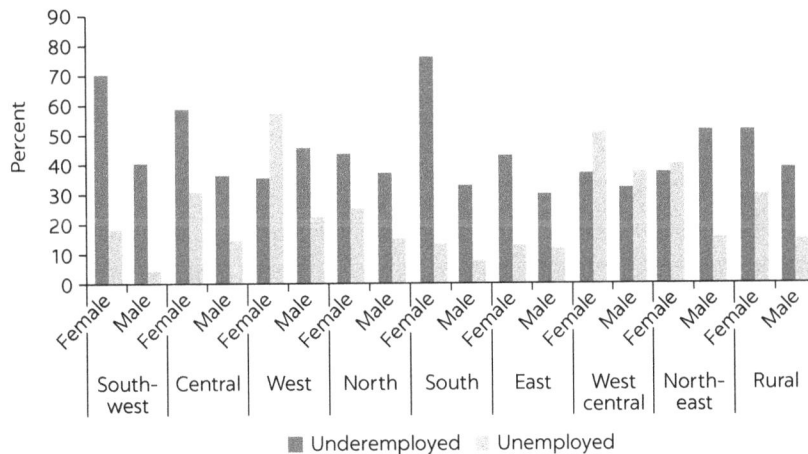

Source: Based on ALCS 2013–14.

rate is highest in the west central region (37 percent), followed by the west (22 percent). In Jawzjan and Paktika provinces, female LFPR is high, and they also have lowest female unemployment rate (map 3A.2).

Policymakers must also pay attention to underemployment, as employment often fails to improve the livelihoods of rural people due to high underemployment. Underemployment prevents female workers from reaching their potential in terms of productivity. Thus, lowering underemployment is crucial to reduce poverty and the vulnerability of rural households. The underemployment rate is about 50 percent for women and about 40 percent for men. It is highest among men in the northeast (52 percent), followed by the west (46 percent) and southwest (41 percent). Livelihoods in these regions are predominantly in agriculture, and subsistence activities are often insufficient to keep farmers employed full time. The female underemployment rate is highest in the south (76 percent),

FIGURE 3.16

Employment by sector in 2013–14: male versus female workers

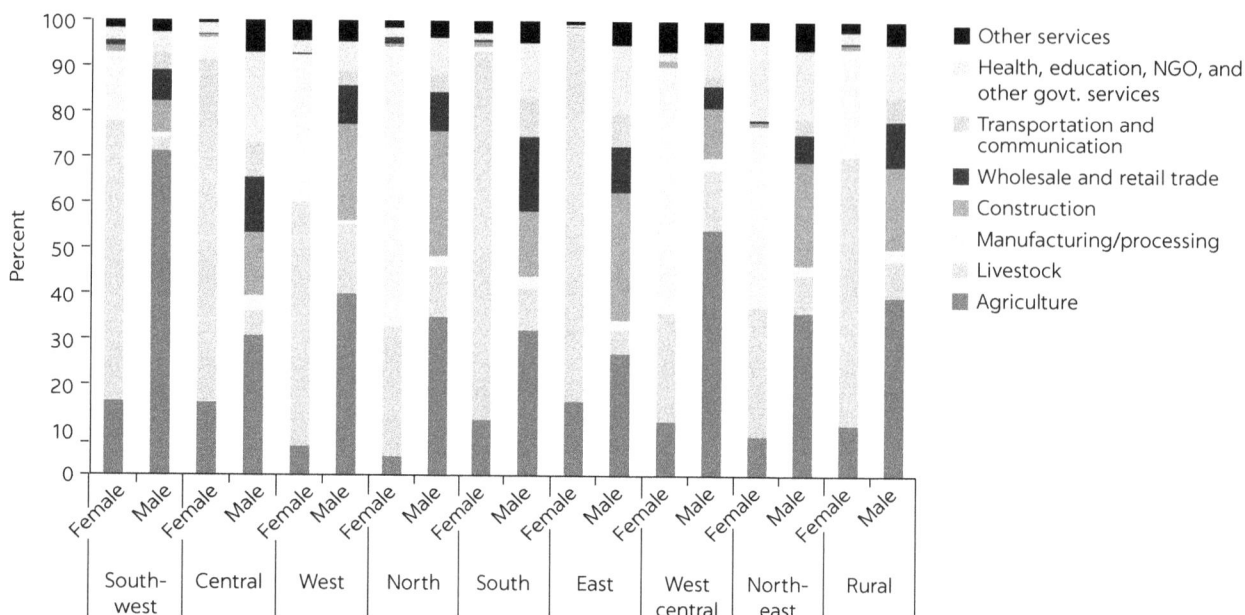

Source: Based on ALCS 2013–14.
Note: NGO = nongovernmental organization.

BOX 3.2

Spotlight: Women's economic empowerment in Afghanistan

Status of women. Women's economic empowerment and participation are increasingly being acknowledged as integral parts of development and sustainability. A forthcoming World Bank study shows stakeholders led many interventions to further women's economic empowerment during the last decade. However, these interventions were often disjointed, uncoordinated, and focused on limited outputs (such as the number of trained women) rather than impacts on empowerment. These findings relate to the absence of a clear definition and contextualization of women's economic empowerment, which has resulted in it being described strictly in terms of employment status and earnings, sidelining interventions to bridge gender gaps in endowments, agency, and economic opportunities. The study shows that literacy continues to be a barrier to women's economic empowerment and that related programs should follow an integrated format to combine skills and literacy for greater impact. In addition, the study highlights that women's economic empowerment starts in the social sphere, where interventions, such as those that facilitate public spaces for women to

gather and build social bonds, are the starting point to support a pathway toward economic empowerment.

Advancing women's economic empowerment cannot take place without improved cooperation and coordination among multilateral and bilateral donors for effective use of resources, building consensus on supporting evidence-based programs with demonstrated innovative approaches, and creating a cohesive approach to tracking program impact. This process cannot happen without improved data collection for gender statistics and making this data publicly available for donors and project implementers for enhanced monitoring and evaluation of progress.

A comprehensive approach: the Women's Economic Empowerment National Priority Program. In response to the needs of women, the Afghan government launched the Women's Economic Empowerment National Priority Program on March 8, 2017. It was designed to operate in all 34 provinces and provide leadership on best practices and rigorous monitoring to deliver tangible results in the form of improved outcomes for women.

continued

Box 3.2, *continued*

Grounded in the country's constitutional guarantee of equal rights for men and women, it supports women's economic participation to increase their agency, autonomy, and well-being. Through its four productive pillars, the program will provide support to women-owned businesses, development of technical and soft skills, investment in creative industries, and inclusive access to finance. These investments will complement and be delivered through existing mechanisms and institutions, focusing on scaling up and coordination of successful interventions.

The program also improves the economic enabling environment through two supporting pillars that will address legal barriers to women's economic participation and improve the quality and use of gender statistics for planning and monitoring women's economic progress. The program is mindful of the need to involve men throughout its design and management to ensure full ownership and support, and lessen the risks of backlash. Islamic scholars, traditional leaders, and community development councils provide guidance and support.

followed by the southwest (70 percent) and the central regions (59 percent). Policies and programs should be implemented not only to create new jobs, but to improve the quality and sustainability of jobs for the underemployed workforce, especially women.

The sectoral distribution of male and female workers is distinct. About 40 percent of the total male labor force is employed in the agriculture sector, while 60 percent of the total female labor force is employed in the livestock sector (figure 3.16). Three of five female workers were employed in the livestock sector in 2014. Chapter 2 discussed the decline of the sector's income share and the low market participation of livestock producers, resulting in a low return for female workers in the sector.

Another striking feature of rural employment is that only about 2 percent of male workers are involved in the manufacturing and processing sector, compared with around 23 percent of female workers. This sector involves agricultural forward linkages and handicrafts, mostly done by rural women in households. Therefore, improving the value chain of the products produced by rural manufacturing and processing may support the rural female workforce by improving livelihoods and creating sustainable employment.

Very few women participate in the formal employment sector, which employs only 2.3 percent of rural female workers, compared with 11.7 percent of male workers. The female workforce is lagging in schooling, skills, and training; the Afghan government and donors working for reconstruction should increase employment opportunities for rural women in the formal sector and strive to improve their education and other skills to make them more competitive in the labor market.

Figure 3.16 also shows a varied sectoral composition of the male and female workforce across rural areas. Agricultural employment among male workers is highest in the southwest (71 percent), followed by the west central region (54 percent) and the west (40 percent); for female workers, it is highest in the southwest, central, and east regions, where about 16 percent are involved in agriculture. Female workers' involvement in the livestock sector is predominant in most regions, reaching more than 80 percent in the south and east, 75 percent in the central region, and 61 percent in the southwest. The manufacturing and processing sector's share of employment among the female workforce is high in the north, where three of five women are involved in the sector, as well as the west central, northeast, and west regions.

The pattern of employment type for male and female workforces in rural areas is also distinct (figure 3.17). Four of every five female workers are unpaid family workers, compared with one of every five male workers. The proportion of unpaid female workers varies; it is higher in the agriculture sector, dominated by the south (95 percent), and lower in the north (45 percent). Salaried female workers account for a much higher share in the northeast than any other region. Self-employment among female workers is high in the north, northeast, and west. The share of manufacturing and processing employment among women is much higher in these regions, implying that most women working in this sector are self-employed. The conclusion is that female workers are much less gainfully employed than male workers.

FIGURE 3.17

Type of employment in 2013–14: male versus female workers

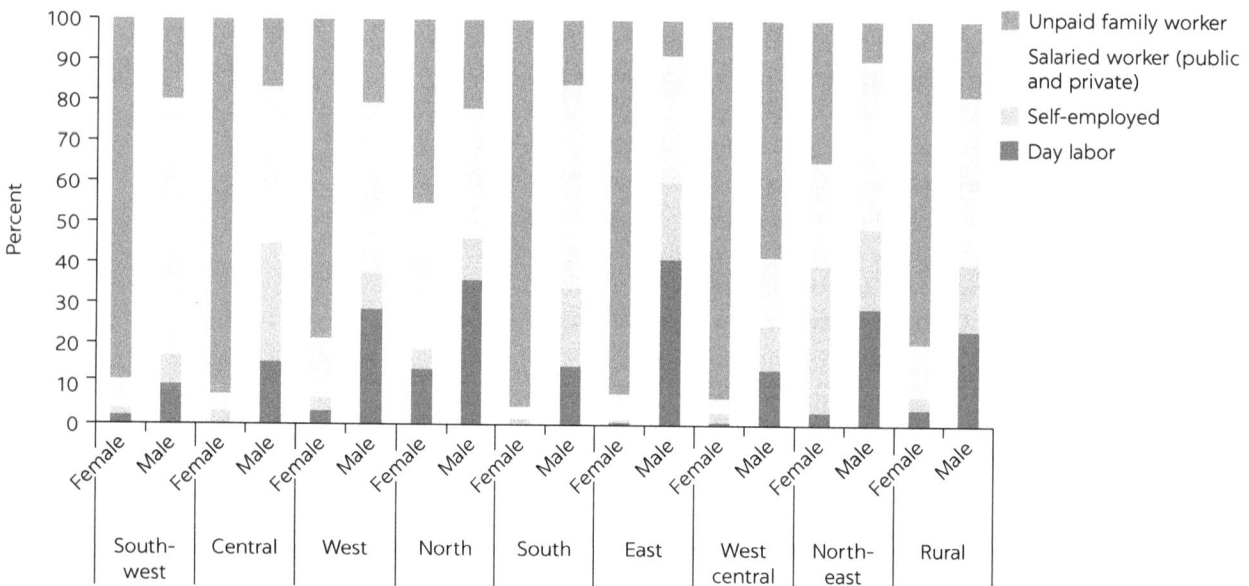

Source: Based on ALCS 2013–14.

BOX 3.3

Spotlight: Promoting women's entrepreneurship and employment through legal reform in Afghanistan

The World Bank Group's *Women, Business and the Law* maps the gender legal differences that affect women's economic empowerment and their ability to get a job and operate a business.[a] The project currently covers 189 economies worldwide, including Afghanistan.[b]

Afghanistan has made progress. For example, it is one of only 22 economies in the project with a quota for women in parliament (27 percent). Quotas can enable a more equitable representation of women in leadership positions and lead to a better reflection of women's interests in decision-making. In fact, at

continued

Box 3.3, *continued*

28 percent, women's representation in the Afghan parliament is the second highest in South Asia, after Nepal.[c] Still, some gender barriers are apparent in the legislation, and an assessment of the current legal and policy framework could be done.

Promoting women's entrepreneurship. Entrepreneurship can be promoted by improving women's inclusive access to finance, which depends on the property available to them as collateral. In Afghanistan, women (unmarried and married) have equal ownership rights to property and can control their property without permission from a husband or a male guardian. However, wives and daughters do not have the same inheritance rights as males. Inheritance rights can be strengthened by procedural provisions. For example, Jordan reformed its laws in 2010, requiring the registration of inherited property and a three-month waiting period during which a woman cannot waive her inheritance rights.[d] This can relieve the potential family pressure on women to give up their inheritance rights.

Married women's access to assets can be strengthened by recognizing their nonmonetary contributions, such as childcare and household responsibilities, as in Indonesia, Malaysia, and Turkey.[e] This would entitle women to an equitable share of the marital property on the dissolution of the marriage, without the need for a direct financial contribution.

The establishment of public credit registries or private credit bureaus with low minimum loan thresholds can also encourage women's access to finance. Small borrowers, many of whom are female entrepreneurs, could leverage their good repayment histories. Information from nonbank institutions can also be used to assess borrower creditworthiness to include those who lack traditional banking relationships. For example, in India and Pakistan, microfinance institutions provide information to public credit registries and private credit bureaus.[f] And after a recent reform, the credit registries in the United Arab Emirates are also collecting data on the payment history of utilities (World Bank 2016b, 41).

Another avenue to increase women's access to finance could be to equalize the legal procedures for applying for national ID cards and passports. The 2016 *Women Business and the Law* notes that women are half as likely to borrow from a financial institution where processes for getting national ID cards are different for women and men. In Afghanistan today, ID cards are mandatory only for men, and women have to be accompanied by a male relative or husband to get a passport.

Supporting women's access to the labor market. Childcare and employment options can promote women's access to the labor market. Afghan legislation, the frontrunner in South Asia, gives fathers the option to take 10 days of paid paternity leave. (Only Bhutan and the Maldives also allow paternity leave, though they grant fewer days.) In addition, Afghan law provides 90 days of paid maternity leave, reducing the burden that childcare puts on women at the expense of their careers. However, as in only 32 percent of the economies that mandate paid maternity leave, the employer bears all the costs; the government is not involved. Hence, from the employer's perspective, the cost of hiring women of reproductive age is higher than the cost of hiring men.

Protecting women from violence. Afghanistan has made major progress in fighting gender-based violence. The 2009 Law on Elimination of Violence against Women allows for criminal prosecution and penalizes child marriage. In addition, unlike in other economies in the region (Bangladesh, India, and Sri Lanka), perpetrators of rape are not exempt from prosecution if they marry the victim. Afghanistan does not, however, have legislation on domestic violence. The 2016 *Women, Business and the Law* highlights that women's life expectancy is likely to be higher in countries where they are legally protected from domestic violence.

a. All information was retrieved from *Women, Business and the Law* (2016), accessible at http://wbl.worldbank.org. The website also provides recent updates to the report's data.

b. According to the *Women, Business and the Law* methodology, the legal analysis focuses on the laws and regulations in Kabul. The data cover statutory law and only applicable religious or personal laws if they are codified.

c. http://data.worldbank.org/indicator/SG.GEN.PARL.ZS.

d. Jordan, Personal Status Law No. 36 of 2010.

e. Indonesia, Marriage Law, Art. 35; Malaysia, Islamic Family Law, Art. 58(4)(a); Turkey, Civil Code, Art. 196.

f. http://wbl.worldbank.org/data/exploretopics/building-credit.

Employment patterns of the bottom 40 percent of income earners: job creation for shared prosperity

One of the World Bank's goals is to achieve shared prosperity by 2030. The availability of employment opportunities and the returns to labor are key, especially when landlessness is acute and labor is the only asset poor people have. Therefore, this study also explores the dynamics and structure of rural employment in Afghanistan by household wealth status. It uses quintiles of asset index (constructed and used by the Poverty Global Practice in poverty analysis for Afghanistan; Balcazar 2016) to study employment patterns of the bottom 40 percent of income earners. This analysis will clarify the labor market preferences of rural households based on wealth status.

The LFPR of the working-age population from the households in the bottom 40 percent of income earners is modestly higher than the LFPR of working-age people from households in the top quintiles (figure 3.18). The underemployment and unemployment rates are also high among workers from households in the bottom 40 percent. In 2014, the underemployment and unemployment rates of workers from the richest quintile were 38 percent and 15 percent, respectively; the corresponding figures for workers from the poorest quintile were 42 percent and 26 percent.

Workers from the poorest households are generally involved in the primary sectors, such as agriculture and livestock, while workers from the richest households are more engaged in the nonfarm sector (figure 3.19). In 2014, about 56 percent of workers from the lowest quintile were involved in the agriculture and livestock sectors, compared to about 42 percent of workers from the highest quintile. Other than agriculture, workers from the richest quintile were involved more in public and private services, wholesale and retail trade, manufacturing and processing, and construction. Of nonfarm activities, workers from the poorest quintiles were involved more in manufacturing and processing. The most important employment status for the poorest households is unpaid family workers, followed by day labor and self-employment; self-employment, unpaid family workers, and salaried work are the most important for the richest households (figure 3.20).

FIGURE 3.18

Labor force participation rate, underemployment, and unemployment in 2013–14, by asset quintile

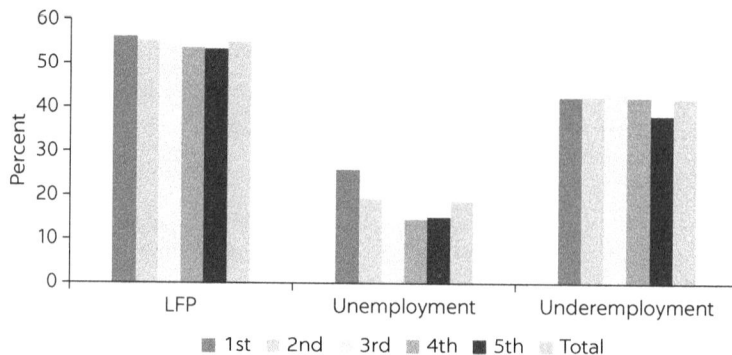

Source: Based on ALCS 2013–14.

FIGURE 3.19

Sector of employment in 2013–14, by asset quintile

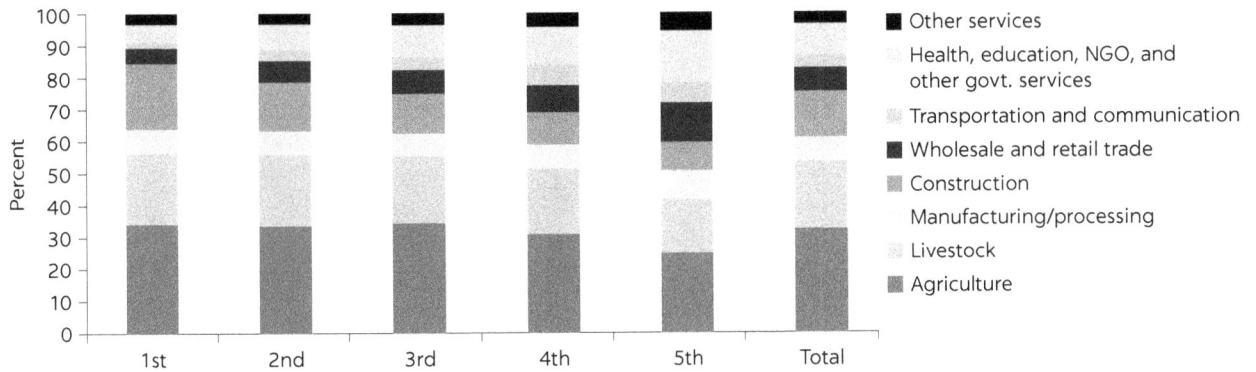

Source: Based on ALCS 2013–14.
Note: NGO = nongovernmental organization.

FIGURE 3.20

Type of employment in 2013–14, by asset quintile

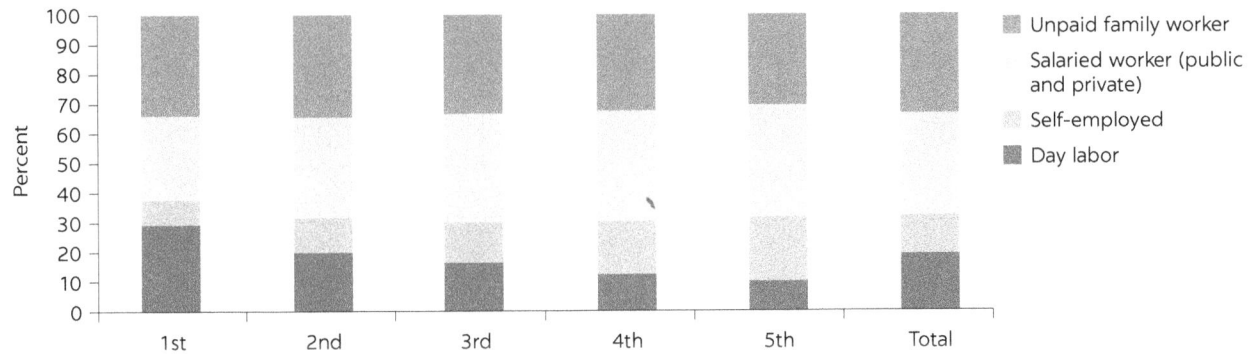

Source: Based on ALCS 2013–14.

ANNEX 3A

MAP 3A.1

Labor force participation rates in 2013–14, male versus female

a. Male LFP rate

b. Female LFP rate

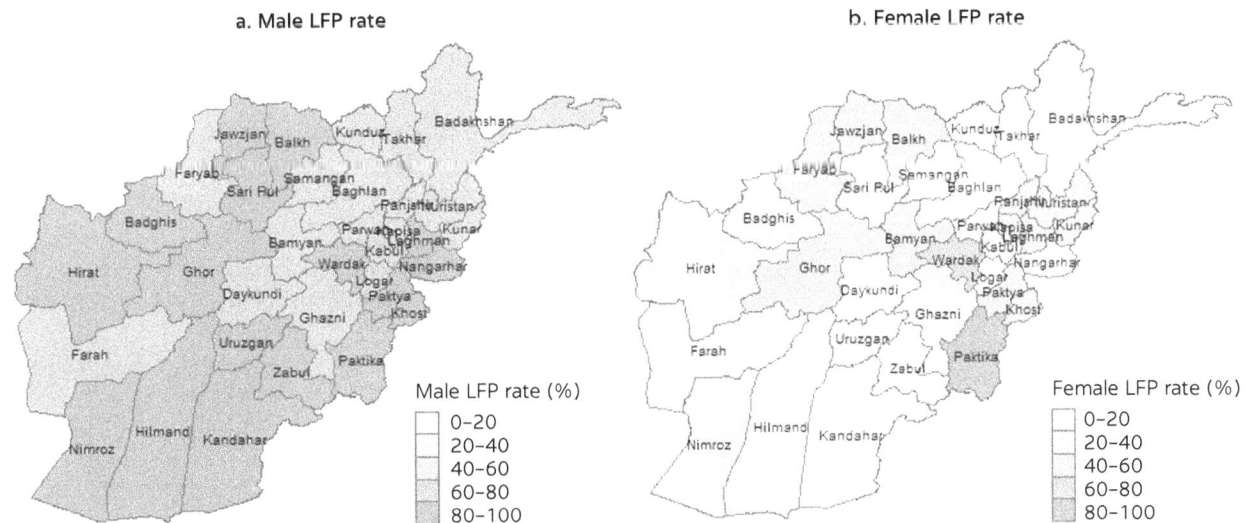

Source: Based on ALCS 2013–14.

MAP 3A.2

MAP 3A.2

Unemployment and underemployment rates in 2013–14, male versus female

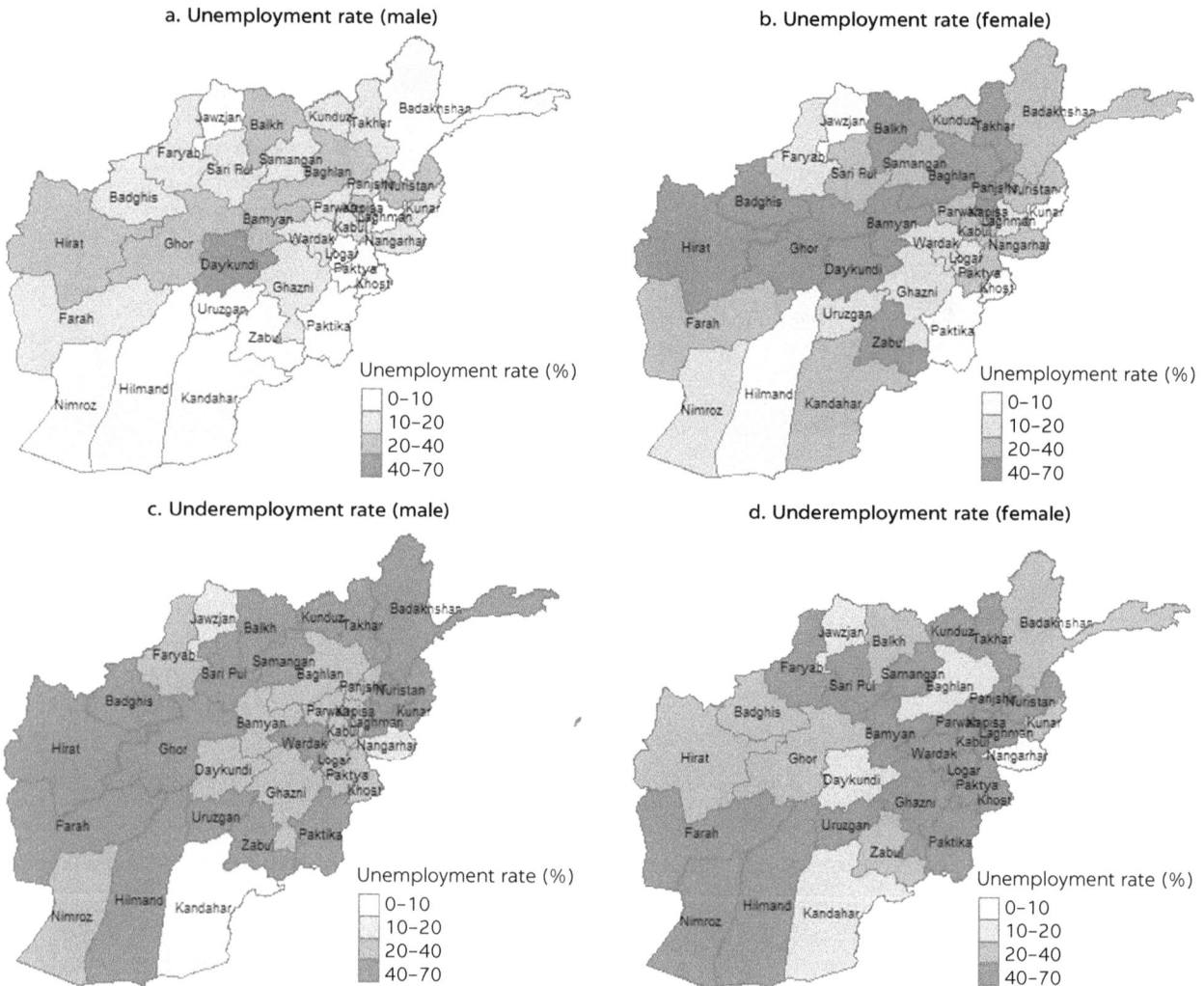

a. Unemployment rate (male)

b. Unemployment rate (female)

c. Underemployment rate (male)

d. Underemployment rate (female)

Source: Based on ALCS 2013–14.

REFERENCES

ADB (Asian Development Bank). 2016. "Asian Development Outlook 2016—Update." Asian Development Bank, Manila.

Balcazar, C. F. 2016. "Notes on an Asset Index for Afghanistan." Poverty GP, World Bank, Washington, DC.

Byrd, W., and D. Mansfield. 2014. *Afghanistan's Opium Economy: An Agriculture, Livelihoods and Governance Perspective.* A Report Prepared for the World Bank Afghanistan Agriculture Sector Review, World Bank, Washington, DC.

Deichmann, U., F. Shilpi, and R. Vakis. 2008. "Spatial Specialization and Farm-Nonfarm Linkages." Policy Research Working Paper 4611, World Bank, Washington, DC.

Gautam, M., and R. Faruqee. 2016. *Dynamics of Rural Growth in Bangladesh: Sustaining Poverty Reduction.* Washington, DC: World Bank.

Hogg, R., C. Nassif, C. G. Osorio, W. Byrd, and A. Beath. 2013. "Afghanistan in Transition: Looking Beyond 2014." World Bank, Washington, DC.

Khandker, S., and H. Samad. 2014. "Non-Farm Drivers of Rural Growth: A Case Study of Bangladesh." Background paper for Dynamics of Rural Growth in Bangladesh, Agriculture Global Practice, World Bank, Washington, DC.

Lanjouw, J., and P. Lanjouw. 2001. "Rural Non-Farm Employment: Issues and Evidence from Developing Countries." *Agricultural Economics* 26 (1): 1–24.

Mansfield, D., and A. Pain. 2007. "Developing Evidence-Based Policy: Understanding Changing Levels of Opium Poppy Cultivation in Afghanistan." AREU Briefing Paper (November), Afghanistan Research and Evaluation Unit, Kabul.

Pain, A. 2012. "Afghanistan's Opium Poppy Economy." Middle East Institute, Washington, DC (accessed September 30, 2016), http://www.mei.edu/content/afghanistans-opium-poppy-economy.

Sen, B., et al. 2014. "Regional Inequality in Bangladesh: Re-Visiting the East-West Divide." BIDS-REF Discussion Paper, Bangladesh Institute of Development Studies (BIDS), Dhaka.

Sen, B., M. Ahmed, and M. Gautam. 2015. "Waves of Change: What Explains Ascent, Descent, and Persistence in Poverty in Rural Bangladesh?" Background paper for Dynamics of Rural Growth in Bangladesh, Agriculture Global Practice, World Bank, Washington, DC.

Shilpi, F., and S. Emran. 2015. "Agricultural Productivity and Non-Farm Employment: Evidence from Bangladesh." Draft, Columbia University, New York.

SIGAR. 2014. "Quarterly Report to the United States Congress." Special Inspector General for Afghanistan Reconstruction, SIGAR, Washington, DC.

UNODC (United Nations Office on Drugs and Crime). 2015. "Afghanistan Opium Survey 2015." UNODC, Vienna.

Ward, C., et al. 2008. "Afghanistan: Economic Incentives and Development Initiatives to Reduce Opium Production." World Bank, Washington, DC. https://openknowledge.worldbank.org/handle/10986/6272.

Wily, L. A. 2004a. "Putting Rural Land Registration in Perspective: The Afghanistan Case." Afghanistan Research and Evaluation Unit, Kabul.

——. 2004b. "Looking for Peace on the Pastures: Rural Land Relations in Afghanistan." Afghanistan Research and Evaluation Unit, Kabul.

——. 2013. "Land, People, and the State in Afghanistan: 2002–2012." Afghanistan Research and Evaluation Unit (AREU), February, Afghanistan Research and Evaluation Unit, Kabul.

Wily, L. A., and World Food Program. "Socio-Economic Profile." Population and Demography, Afghanistan.

World Bank. 2005. "Afghanistan: Country Gender Assessment, National Reconstruction and Poverty Reduction, the Role of Women in Afghanistan's Future." World Bank, Washington, DC.

——. 2014. "Women's Role in Afghanistan's Future: Taking Stock of Achievements and Continued Challenges." World Bank, Washington, DC.

——. 2015. "Afghanistan: Poverty Status Update." World Bank, Washington, DC.

——. 2016a. "Fragility and Population Movement in Afghanistan." World Bank-UNHCR Policy Brief, World Bank, Washington, DC.

——. 2016b. "2016 Women, Business and the Law Report." World Bank, Washington, DC.

——. 2017. *Land Governance Assessment Framework (LGAF) Afghanistan.* Washington, DC: World Bank.

——. Forthcoming. "Mapping and Lessons Learned of Women's Economic Empowerment." World Bank, Washington, DC.

4 Evidence-Based Practice Recommendations for Jobs in Agriculture

LESSONS FROM THE PUBLIC AND PRIVATE SECTORS

INTRODUCTION

To achieve poverty reduction and shared prosperity, the rural economy needs to overcome the challenges discussed in chapters 1–3 (lack of sufficient irrigation and agricultural diversification, a weak nonfarm economy, and poor access to credit and markets) and create more, inclusive, and sustainable jobs. The Afghan government, the World Bank, and other donor agencies initiated numerous reconstruction programs in the aftermath of the fall of the Taliban. The establishment of the Afghanistan Reconstruction Trust Fund (ARTF) was aimed to, amongst others, improve agricultural productivity, rural livelihoods, and the functioning of rural markets. The fund is the main channel through which donor agencies and partner countries can support reconstruction programs and policies aligned with the Afghanistan National Development Strategy and the National Priority Programs. Between 2002 and 2016, donors contributed more than $9 billion for service delivery and reconstruction across Afghanistan; however, overall economic conditions have deteriorated in recent years, and poverty and unemployment in rural areas continue to be more pervasive and persistent than in urban areas.

Against this backdrop, this chapter uses an agricultural jobs lens to explore the roles and performance of major agriculture and rural development programs that are supported by the World Bank. It also discusses the role of the private sector in supporting job creation and linking farmers to markets. How many full-time equivalent (FTE) jobs do these programs create? What types of jobs are created: direct, indirect, or induced? Are the jobs seasonal, temporary, or sustainable? Are projects' job effects significantly addressing local unemployment and underemployment in rural areas? This chapter seeks answers to these questions using information from project documents and from a small sample of private sector business in an agricultural value chain. Based on a mixed-method evaluation of ex-ante data, we can conclude that the agriculture portfolio includes underlying references to the agenda of promoting the creation of more, better, and inclusive jobs. But identifying the causal pathway for and quantification of jobs creation in agricultural projects remains a challenge.

PARTNERSHIPS AND DONOR COORDINATION IN THE AGRICULTURE SECTOR

Afghanistan still relies on international assistance for job creation, service delivery, and strengthening the security environment. With international assistance gradually declining and domestic revenue mobilization falling short, the fiscal situation threatens the foundations of the country's development agenda. With the ARTF, the government and donor agencies are supporting reconstruction initiatives to improve the livelihoods of the rural poor. While programs such as the National Solidarity Program III[1] (NSP III) generated "more" jobs for the rural poor by creating short-term work opportunities, the National Horticulture and Livestock Project (NHLP) has been generating "sustainable" jobs through the development of livestock and horticulture.

The Afghan government and the United Nations Assistance Mission in Afghanistan lead donor coordination. The World Bank Group plays an important role in convening donors around a common agenda, and engages ARTF donors at the technical level through the fund's Strategy Group. In agriculture, for instance, this common agenda is aligned with the country's National Agricultural Development Framework, a comprehensive plan structured around development priorities to improve productivity and resilience in the sector through improved access to key inputs and agricultural extension services, enhanced agribusiness opportunities, and improved land and water management, including expanding the area under irrigation.

As of June, 2016, the agricultural portfolio of selected donors over the past 12 years indicates that investments go toward six types of interventions: agricultural production and marketing; irrigation; community-driven development (CDD); capacity building; food security; and environment management. The main donors include the United States Agency for International Development (USAID), the Asian Development Bank, the UK Department for International Development, Australia, Canada, and the European Union, which together account for about 88 percent of the total portfolio (tables 4A.1 and 4A.2). Agricultural production and marketing projects account for nearly 80 percent of the total budget, followed by irrigation (15 percent), CDD (3 percent), and capacity building (2 percent). Investments in food security and environmental management are negligible. USAID financed most investments in capacity building (61 percent) and agricultural production and marketing (59 percent); the Asian Development Bank financed most investments in irrigation (59 percent).

Interventions and priority setting: the role of world bank projects

The World Bank's agriculture and rural sector investment operations in Afghanistan can be divided into four main types of interventions: agricultural production and marketing; irrigation and on-farm water management (OFWM); rural enterprise development; and rural livelihoods and CDD (box 4A.1). It is important to understand the linkages toward more, better, and inclusive jobs resulting from these "standard" intervention types. There are design and supervision implications for operations in Afghanistan, as well as in other countries where similar work is being carried out.

Evidence from these types of projects suggests that the development of community-based enterprises and integrated value chains in rural areas, improved access to services and resources via non-governmental organizations (NGOs) and government agencies, improved technologies in livestock and orchards, and efficient water use are able to support more, sustainable, and inclusive jobs. This evidenced-based review includes projects under the agriculture portfolio in Afghanistan that have the strongest implications for job creation. Admittedly, project design did not always aim to create jobs, which necessitated applying a "qualitative matching" method to classify project components according to intervention type. Thus, project documents and results frameworks were the primary sources used to identify the jobs outcome.

While the ultimate goals of World Bank agriculture projects in Afghanistan are to achieve poverty reduction and shared prosperity in a fragile and conflict setting, they are designed and implemented to achieve these goals by improving the livelihoods of the rural poor and farmers. Therefore, the projects in this analysis resulted in a sizable increase in job creation. For example, the aim of the OFWM project is to improve farmers' water efficiency and agricultural productivity, but it has also created many short-term jobs in rural areas in 11 provinces and is expected to support better and sustainable jobs by reducing underemployment among beneficiary farmers. Thus, the project creates more and sustainable jobs for the rural poor and rural farmers. While OFWM's indirect job creation impacts were projected to be higher, it has generated about 4,500 direct FTE jobs, many of them short-term and related to infrastructure.

Under the framework of job creation through rural enterprise development, the Afghanistan Rural Enterprise Development Program (AREDP) generates more, sustainable, and inclusive jobs for rural women and the unemployed through the provision of access to finance and skill training. Most of the jobs are generated downstream of the agricultural value chain, and a major share is generated for women. The program has created about 29,000 jobs through loans, 8,000 through enterprise groups, and 20,000 through the development of small and medium-sized enterprises (SMEs) in five provinces.

NHLP, under the livelihood framework of agricultural production and marketing, creates more, sustainable, and inclusive jobs for rural people. The previous chapters demonstrated that most women in the labor force, as well as a large share of youth in rural areas, are employed in the livestock sector. Moreover, livestock and livestock wage labor are important sources of employment for workers from landless households. Our analysis of project documents suggests that activities generate about 10,000 direct and indirect FTE jobs. Although we do not have gender-specific jobs estimates, we do know that about half of beneficiaries are women. About 95,000 women benefit from NHLP activities, compared to 98,000 men. Horticulture extension activities have benefitted about 62,000 women and 78,000 men in 23 provinces. Women benefit most from livestock extension activities (33,000 women versus 20,000 men).

NSP III, which operated in most provinces, mobilized resources and technical assistance for rural people to improve their livelihoods under the CDD approach. While building pathways for long-term sustainable development, NSP III provided financial support to thousands of small infrastructure projects under ARTF initiatives. These projects have helped by generating short-term employment in rural areas, or about 146,000 FTE jobs, of which 137,000 are direct FTE jobs. In addition, NSP III's Maintenance Cash Grant (MCG) Project, as a part of the Afghan government's Jobs for Peace initiative, has generated another 22,000 rural jobs, mostly unskilled positions. Therefore, NSP III's CDD approach has generated more jobs and more inclusive jobs, as most require low or no skills. It also helps generate sustainable jobs indirectly by improving rural infrastructure and gender-balanced institutions, which help connect local producers with national agricultural value chains and catalyze the rural nonfarm economy. As lack of infrastructure is one of the greatest impediments for the underdeveloped private sector, NSP III's rebuilding projects are crucial to increasing private investments and catalyzing the private sector to operate in rural areas. The growth of private sector operations in rural areas will generate sustainable and inclusive jobs.

THE PRIVATE SECTOR'S ROLE IN AGRICULTURE DEVELOPMENT

Agribusiness, whose role in economic transformation in low-income countries is well documented (Da Silva et al. 2009; IFC 2013), can play a central role in Afghanistan's economic recovery and growth. Understanding this role in sustaining jobs in the economy is a key analytical insight in gauging the magnitude of direct and indirect employment effects in the economy.

Agro-processing firms add value to agricultural products, generating employment opportunities and providing wage income. In addition to domestic value additions, their impact is amplified through backward and forward linkages in the economy. Backward linkages with input suppliers (such as family farms, aggregators, and cooperatives) and service providers (such as transporters) further create jobs and income-bearing opportunities locally and in other regions. Similarly, agro-processors contribute to additional job creation and economic spillovers through forward linkages with distributors, wholesalers, and retailers. Robust upstream and downstream linkages with SMEs serve multiple objectives. These linkages provide the lead agro-processing firm with the necessary supply chain architecture to connect agricultural produce to domestic and international consumers. This process also sustains SMEs by creating demand for their products and services, further enhancing their capabilities and market access.

We used a case study approach to estimate the number of indirect jobs created by a successful agro-processor through backward and forward supply-chain linkages in the Afghan economy. Both domestic and foreign investments in the agribusiness sector may yield benefits for agro-processors: They can relax capital constraints on growing firms, enable higher production levels through the acquisition of more and better inputs, facilitate investments in equipment and workers that boost productivity, and facilitate compliance with international quality and safety standards. Among the most fundamental anticipated outcomes of higher levels of investments is the creation of jobs in agro-processing firms. Supply chain linkages of the lead agro-processor, as noted earlier, further job creation in other segments of the supply chain. This indirect job creation can be significant and may translate into job creation in rural and remote areas, leading to geographically balanced job creation and economic growth.

ASSUMPTIONS AND RISKS FOR THE EFFECTIVENESS OF DEVELOPMENT INTERVENTIONS IN JOB CREATION

Assumptions

There are four basic assumptions: the absence of extreme weather events and abrupt changes in climatic patterns; a stable security, political, and institutional environment; availability of adequate infrastructure and access to finance; and investments in additional or improved equipment to expand the scope and quality of production.

- Climate and agricultural activities are inextricably linked. As consequences from climate change can reduce or disrupt agricultural activity, the development of interventions, as laid out in the theories of change, is contingent on the absence of such events.
- Safety and political stability can affect local investment decisions and the ease of establishing export networks. Disruption of regional and local markets can prevent the absorption of additional output that interventions aim to generate. Moreover, interventions assume strong and transparent institutional capacity, which is particularly important given the cooperation between the World Bank Group and its development partners and government officials at different levels.
- Adequate infrastructure and access to finance are vital. Because many interventions aim to promote commercialization, adequate transport services and processing infrastructure are necessary to deliver high-quality agricultural output in the timely and reliable way that commercial partners expect.
- Investments in additional or improved equipment to expand the scope and quality of production assume the existence and functioning of a financial system that allows project participants to avoid liquidity constraints that might prevent them from using the improved technologies.

Risks

Risks can be categorized as factors that prevent projects from taking their full effect or mechanisms that lead to unintended consequences in program areas. Despite careful planning, interventions might not achieve their goals and may have unintended consequences for target communities and participants. Therefore, it is essential to be aware of potential risks when designing projects and to consider safeguards to prevent or mitigate unintended consequences. Implementing partners and other stakeholders should also plan and set their expectations in accordance with circumstances.

Creating more, better, and inclusive jobs requires a multisectoral approach. Concentrating on only one sector risks overlooking the extent to which interventions in other sectors contribute to the jobs agenda. For example, agriculture and rural development interventions are just one aspect of addressing worker productivity gains. Improving infrastructure and access to finance, and disseminating knowledge on improved agricultural techniques and business skills, are also essential to achieving intended outcomes.

Time is another critical factor that might prevent projects from achieving their intended jobs outcomes. Realistically, projects may not be able to create more, better, and inclusive employment in a three- to five-year implementation period. It is helpful to identify feasible goals that set target communities on track to achieve jobs results within the project timeframe and plan for post-completion monitoring and evaluation to articulate results.

The lack of data to support the analysis of more, better, and inclusive jobs is another challenge. Many results frameworks prioritize immediate outputs over jobs outcome, as high costs, security issues, and limited capacity can make data collection difficult. With limited data to draw on, drafting precise, realistic results frameworks can be even more challenging.

Another risk is that intensification of agricultural activity might create unintended consequences. For example, increased agricultural activity can lead to negative environmental impacts, such as soil degradation or changes to the balance of ecosystems. Also, incentivizing farmers to increase fertilizer application might lead to overuse—especially if farmers have little or no experience or training in how to use chemical fertilizer—which may cause environmental damage and health hazards if chemical runoff contaminates water sources. Increased child labor is another risk, as intensified cultivation and harvesting might require farming households to rely on the support of all household members.

JOBS FROM PUBLIC SECTOR INTERVENTIONS: EVIDENCE FROM WORLD BANK PROJECTS

Monitoring and evaluation frameworks implicitly monitor job creation; they do not design metrics frameworks that provide a causal pathway to explicitly measure more, sustainable, and inclusive jobs. The project documents and monitoring and evaluation frameworks of programs reviewed in this study did not analyze the influence agriculture and rural development activities had on job results. Projects related to irrigation and OFWM track output indicators relevant to job creation (for example, increase in irrigated areas, number of project beneficiaries, or number of people trained). They do not, however, relate these findings to job results.

Furthermore, the reviewed documents and frameworks seldom referred to the agenda of "sustainable jobs." Although some projects track productivity gains, they measure different types of agricultural productivity (such as land and water) without analyzing worker productivity. Similar to what has been observed for job creation, some projects track indicators relevant to labor productivity without measuring labor productivity explicitly. Another area requiring greater attention is the "sustainability dimension" that uses skills development. For example, job creation results indicate that the NSP III generates many direct jobs, primarily for unskilled rural workers, but most are unsustainable.

When analyzing inclusivity, it is important to distinguish between addressing gender aspects and promoting inclusive jobs for youth, the bottom 40 percent of income earners, and lagging regions. For example, to date, no agriculture and rural development project in Afghanistan has emphasized youth employment in its documents. Because all projects target smallholder farmers or rural communities, where poverty tends to be most acute, these groups do garner some attention. However, none of the reviewed results frameworks featured indicators disaggregated by age or socioeconomic dimensions. On the other hand, gender is the inclusivity component that receives the most attention. The NHLP, for example, analyzed which and what parts of value chains promoted employment for women. All three iterations of the National Solidarity Program address gender inclusivity from the perspective of female participation in local governance institutions, rather than through a "jobs lens." The focus on women is widely reflected in the program's results frameworks, with 80 percent of the projects containing indicators disaggregated by gender.

The agriculture portfolio includes references to promoting the creation of more, sustainable, and inclusive jobs. However, it is important to increase the focus on inclusive jobs and ramping up investment in "more jobs" to reap the full potential of the sector for the labor market. To facilitate future agricultural project design to emphasize these three dimensions of jobs, table 4A.3 provides a list of higher-level indicators that can improve more intense job monitoring. We recommend these be disaggregated by gender, age, and socioeconomic status, where applicable.

Agricultural production and marketing: the role of National Horticulture and Livestock Project

NHLP concentrates on linking rural farm producers with markets by promoting improved production practices for horticulture and livestock producers through the gradual development of farmer-centric service delivery. However, the project lacks a component of linking the producers with the markets.

Background. Horticulture and livestock are key subsectors for inclusive economic growth and sustainable employment in Afghanistan. To reduce poverty and unemployment in rural areas, these subsectors must play a vital role in livelihood improvements and job creation. Much of the agricultural infrastructure was destroyed during three decades of war, including infrastructure essential for fruits and livestock producers to access markets. Afghanistan was the world's top supplier of horticulture products in the 1970s, but its exports today account for only a small portion of global exports. In the 1970s, Afghanistan was also producing a sufficient level of meat and milk for its own consumption, and exporting wool, carpets, and leather products. The good news is that horticulture production has started to improve, and the sector's output has more than doubled in the past decade, from Af 6,689 million in 2006–07 to Af 16,478 million in 2015–16. Livestock production, however, remained stagnant throughout the same period, at Af 11 million–12 million.

To improve productivity and employment in the horticulture and livestock subsectors, Afghanistan's Ministry of Agriculture, Irrigation, and Livestock launched the NHLP in 2013, with support from the World Bank. It replaced and scaled up the Horticulture and Livestock Project, which the ministry implemented from January 2007 to December 2012 with financing from the World Bank and the ARTF. Using a demand-driven approach, NHLP supports and assists farmers and livestock owners to expand their production with improved technologies and practices. Active in more than 100 districts in 23 provinces, the program is scheduled to run for six years, from January 2013 to December 2018.

Key inputs and outcomes. The latest Implementation Status and Results (ISR) report of 2016 shows that, since 2012, NHLP has reached 285,205 farmers and herders, of whom about 42 percent are women. It has achieved or surpassed many of its targets. For example, its goal was to rehabilitate (direct/indirect) orchards in 6,000 hectares, but has already ensured rehabilitation (direct/indirectly) in 70,500 hectares. In addition, it helped rural households establish about 9,000 hectares of new orchards during 2013–15.

NHLP also supports thousands of rural farmers through horticulture and live-stock extension services. During 2013–15, horticulture extension services supported about 140,000 beneficiaries (figure 4.1) and the livestock services supported an additional 60,000 beneficiaries, of whom 63 percent were women. Chapter 3 showed that most female workers in rural areas are employed in the livestock sector; it follows that NHLP's livestock sector support helped women the most.

Map 4.1 shows a modest positive association between the number of NHLP beneficiaries and the intensity of livestock employment in rural areas. Livestock accounts for a high level of employment in many provinces in the project areas. On the other hand, the number of beneficiaries is high in provinces with high unem-ployment rates, which may imply that the project might result in solid employ-ment generation for the unemployed (see figure 4A.3). NHLP activities, however, are limited to 100 districts in 23 provinces; with this scale of operation, it is hard to have significant impact on the overall unemployment situation in rural areas.

National Horticulture and Livestock Project's jobs effects. In addition to direct employment, NHLP has generated many indirect jobs by expanding orchard and livestock production through technical support and financial assistance.

• NHLP horticulture extension activities have created about 10,000 FTE jobs, about 60 percent of which are long-term jobs (table 4.1). Most of these resulted from direct job creation effects. Thus, NHLP's horticulture activi-ties create both more and sustainable jobs.
• Figure 4.1 also shows that, in most provinces, maximum jobs are the outcome of direct job creation, and that these are long-term jobs. The job creation effects NHLP's horticulture activities are high in Balkh, Kabul, and Samangam provinces.

FIGURE 4.1

Beneficiaries of National Horticulture and Livestock Project

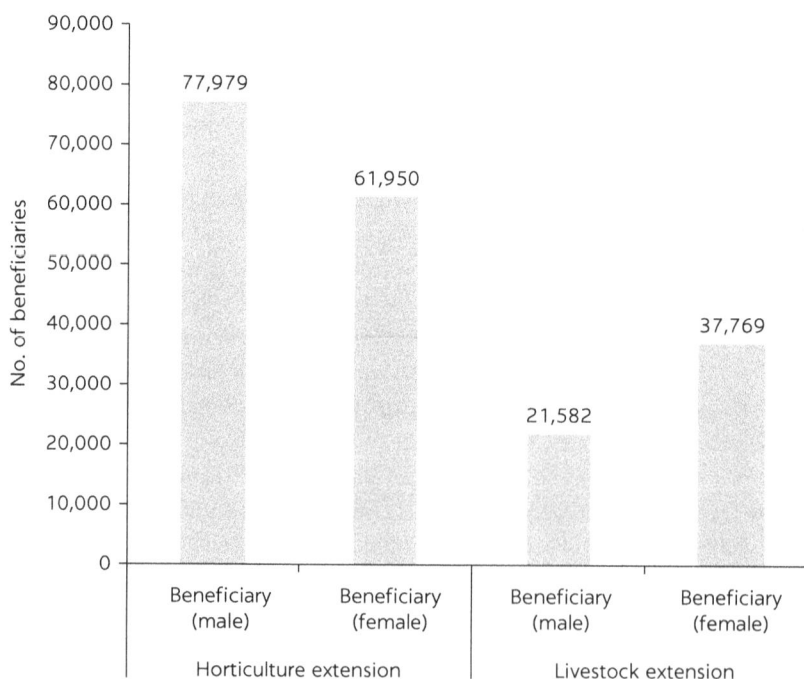

Source: Project documents 2016.

MAP 4.1

Intensity of livestock employment, and National Horticulture and Livestock Project beneficiaries, as of Sept. 2016

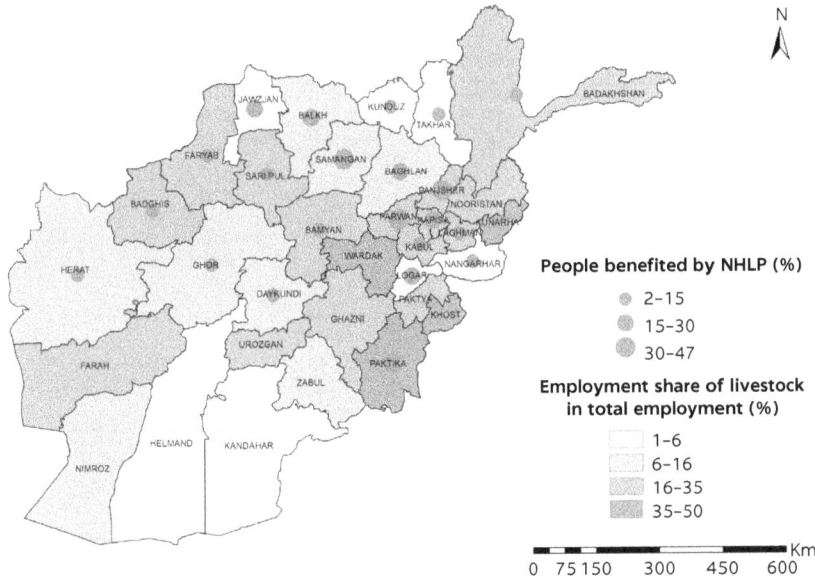

Source: Project documents 2016 and ALCS 2013–14.

TABLE 4.1 **Job creation by National Horticulture and Livestock Project (horticulture)**

	2015	2016
Project staff (number)	899	1,088
Direct employment (no. of long-term, FTE jobs)	6,269	6,041
Direct employment (no. of short-term, FTE jobs)	735	630
Indirect employment (no. of short-term, FTE jobs)	2,181	2,299
Total jobs creation	10,084	10,058

Source: Project documents 2016.
Note: FTE = full-time equivalent.

- While it is believed that NHLP's livestock extension activities have important job creation effects, we do not have information about the actual number of labor days they have generated. It is estimated that more than 50,000 farmers are benefitting from animal health and extension activities. NHLP's livestock extension services play an important role in improving workers' labor productivity, including many female and youth, and generating more jobs. Table 4.2 shows that about 355,000 FTE equivalent jobs are generated through livestock and animal extension services.[2]

Though NHLP plays key roles in improving livelihoods and employment through development of livestock and horticulture, its geographic coverage and numbers of beneficiaries are low. To improve livelihood conditions and generate employment in these subsectors, activities and geographic coverage need to be significantly scaled up. (This is underway through an additional finance phase.) Moreover, a component to support linking producers with markets can be developed in the next phase to help realize the subsectors' full potential.

TABLE 4.2 **Jobs created and additional revenue generated through animal health and extension, 2015**

PROVINCE	BENEFICIARY (NUMBER OF FARMERS)	TOTAL FTE JOBS	TOTAL INCREMENTAL REVENUE (AF)	INCOME PER FTE JOB (AF)
Badakhshan	111,874	18,182	303,871,309	16,713
Badghis	94,815	15,409	240,568,515	15,613
Balkh	157,230	25,552	190,436,110	7,453
Bamyan	54,169	8,804	105,715,237	12,007
Faryab	135,537	22,026	267,661,480	12,152
Ghazni	97,471	15,840	116,303,613	7,342
Helmand	119,188	19,369	336,971,149	17,397
Herat	332,893	54,097	411,104,757	7,599
Jawzjan	96,796	15,731	158,875,241	10,100
Kabul	70,840	11,513	64,767,831	5,626
Kandahar	86,206	14,010	263,158,792	18,784
Kapisa	52,059	8,461	93,789,082	11,085
Khost	39,219	6,374	114,999,742	18,042
Kuner	52,661	8,558	177,161,692	20,700
Kunduz	63,777	10,365	164,650,996	15,885
Laghamn	49,839	8,100	135,162,283	16,687
Logar	51,580	8,383	40,088,668	4,782
Nangarhar	136,804	22,232	236,092,554	10,619
Paktia	33,068	5,375	58,216,636	10,832
Panjshir	22,345	3,632	71,984,344	19,820
Parwan	88,993	14,463	118,009,594	8,160
Takhar	151,157	24,565	223,635,634	9,104
Wardak	87,067	14,150	58,125,026	4,108
Total	2,185,588	355,189	3,951,350,284	11,125

Source: Project documents 2016.

Irrigation: the role of the On-Farm Water Management Project

Improvement of irrigation and OFWM has vast potential for increasing agricultural productivity, improving water efficiency, and bringing new areas under irrigated agriculture. Initiatives to improve irrigation facilities and OFWM have significant potential to create more, sustainable, and inclusive jobs in rural Afghanistan, where most people rely on agriculture for their livelihoods and food security. Programs for increasing irrigation facilities can create direct short-term jobs, which are generally inclusive, as well as indirect long-term sustainable jobs by improving farmers' land productivity. Irrigation and OFWM investments can also reduce the underemployment rate among farmers and generate new jobs.

Background. With World Bank support, the Ministry of Agriculture, Irrigation, and Livestock implements the $41 million OFWM project in 46 districts in 11 provinces. While its broad objectives are to increase farm households' agricultural productivity and income, its specific objectives are to raise land productivity (wheat yield, ton/ha), water productivity (wheat yield for each

TABLE 4.3 **On-farm water management job creation**

COST ITEMS	SHARE OF COST (%)	ANNUAL AVG. NO. OF LABOR DAYS FOR IRRIGATION REHABILITATION PROJECT (400 HA/SCHEME)	COMMENTS
Labor (project cost)	20	4,500	
Skilled labor (labor cost)	40	1,500	100% of skilled workers are male
Unskilled labor (labor cost)	60	3,000	100% of unskilled workers are male
Key and administrative staff (project cost)	6	10	Key and administrative staff: 90% male, 10% female

Source: Project documents 2016.

volume of water used), and irrigated area through a more efficient irrigation system. OFWM began in 2011 and is expected to close in 2019.

Productivity of agricultural crops in project areas increased by 30 percent, and water use efficiency improved by 25 percent. OFWM sought to increase water productivity from 0.63 in 2011 to 0.75 in 2019, but had reached 0.76 by the time of this analysis. Irrigated land has increased 5 percent, short of the targeted 15 percent increase from the 2011 base level (ISR 2016). The ISR also showed that OFWM had increased land productivity and farmers' incomes in project areas.

Job creation effects. OFWM works to reduce severe water loss and its velocity in the irrigation system. Therefore, its main role, in terms of job creation, is to create indirect jobs by raising the work hours of underemployed farmers and to induce further employment through an increase in farm household income. Although the project does not have much job-related information, table 4.3 provides some ideas about the potential for direct job creation. It shows that OFWM created about 4,500 direct jobs, of which two-thirds were unskilled workers, mostly males. Most are unsustainable short-term jobs, but the project does have the potential to create sustainable jobs through indirect job creation. However, due to lack of data, the measurement of such indirect job creation effects is beyond this case study.

Although OFWM has been successful in improving farmers' agricultural productivity and efficient use of water, its job creation effects remain unclear due to lack of data. To understand the actual job creation effects, a measurement of employment effects must be incorporated into the project ISR. Moreover, to improve overall agricultural productivity and rural livelihoods, project coverage needs to be expanded, which the additional finance phase is facilitating.

Rural enterprises: the role of Afghanistan Rural Enterprise Development Project

Rural enterprises are integral to rural livelihoods and a key source of employment, and their importance as a driver of rural economic growth and poverty reduction is growing in the developing world. Thus, fostering rural enterprises can be key for rural growth and sustainable development in Afghanistan, where expansion can generate more, better, and inclusive jobs for the rural poor. The World Bank launched AREDP to support rural enterprises through increased access to finance, technical knowledge, and business support services. AREDP has created enhanced opportunities for employment and incomes for rural people, particularly women.

Background. Funded by the World Bank and other donor agencies, AREDP is the flagship program of Afghanistan's Ministry of Rural Rehabilitation and Development (MRRD). The implementation period is 2010–18 at total cost of $92 million, of which the World Bank is contributing $35 million. The principal objective is to improve employment opportunities and income for rural people (male and female) through the development of community-based enterprises and integrated value chains. AREDP's goal is to create more jobs and income by providing support to local businesses in market access and finance. It aims to achieve the following:

- Support local businesses by providing business services in market orientation, sustainable businesses, facilitating client decisions, sharing best practices, and vertical integration

- Provide knowledge-based and financial services to community-based rural enterprises and SMEs that provide business advisory and financial services to rural SMEs

- Support enterprise development activities with marginalized rural communities, such as Kuchis (nomadic pastoralists), and people with disabilities

AREDP is active in 694 of 2,365 community development councils (CDCs) in 24 of 67 districts in five provinces: Balkh, Bamyan, Herat, Nangarhar, and Parwan (MRRD 2016). These provinces are home to about 31 percent of the total rural population (Central Statistical Office [CSO] 2014); in total, activities cover about 41 percent of the provinces' total population. AREDP works in only four of 15 districts in Balkh, which account for 20 percent of the province's population (CSO 2014). In Bamyan province, where the Hazara ethnic group is prominent, AREDP is active in four of seven districts, an area that is home to 64 percent of the population. AREDP is present in only four districts in Herat, covering 140 villages that account for about 32 percent of the population. In Kandahar, it operates in four of 18 districts that account for 60 percent of the population. Six project districts in Nangarhar account for 40 percent of the population. AREDP covers five districts and 149 villages in Parwan, accounting for 57 percent of the province's population.

Key inputs and outcomes. AREDP is one of the most successful programs in Afghanistan. As of August 2016, it had achieved the following outcomes[3]:

- Established 5,260 savings groups;
- Established 1,360 enterprise groups;
- Supported 593 SMEs;
- Established 493 village savings and loan associations;
- Supported 1,304 people with disabilities and their families;
- Supported 1,258 Kuchis and their family members;
- Achieved total savings of Af 169.3 million;
- Disbursed loans valued at Af 255 million and recovered loans valued at Af 132 million, with 26,566 total borrowers (male and female).

Job creation effects. Although AREDP is expected to have a large induced job creation effect, estimating it is beyond scope of this analysis. Direct and indirect job creation could also provide insights about the project's job creation potential. AREDP forms savings groups and village savings and loan associations to support local people to improve access to finance for new entrepreneurs, and we measured job creation from such lending activities. When financial institutions finance or lend money to a person (borrower) to start, grow, or expand a business over an agreed-upon repayment period, this loan was considered as "one employment generated" (as one loan means one employment generated). Table 4.4 and table 4A.5 in annex 4A provide a glimpse of AREDP's job creation through loans.

The table shows that about half of the jobs created through lending activities were in the livestock and poultry sectors. It also reveals that more women were involved in the livestock, poultry, handicrafts, small business, and small machinery subsectors than men. More jobs for men were generated in agriculture and shop keeping. Furthermore, the average loan amount was higher for men than women, implying that even small loans can create more jobs for women. For example, the average loan amount for a woman in the livestock sector was Af 6,584; it was about Af 9,739 for each man.

Job creation is highest in Nangarhar province, followed by Balkh and Herat. AREDP provides support through loans to livestock producers in the five provinces where it operates. Job creation in the handicrafts sector is relatively high in Bamyan province. AREDP also creates direct, indirect, or seasonal[4] jobs by supporting local enterprises (tables 4.5 and 4A.6).

Our research of project support to local enterprises revealed the following:

- About two-thirds of the jobs AREDP creates are through direct job creation, much more than indirect and seasonal jobs;
- AREDP creates the most direct and seasonal employment in Nangarhar province;
- Income growth from 2015 to 2016 was higher for women than men;
- Job creation was highest in aviculture, dairy, handicrafts, and food processing;
- Although job creation for women was higher than for men, women, on average, earned much less.

TABLE 4.4 **Afghanistan Rural Enterprise Development Program employment generation through loans by gender**

SECTOR	EMPLOYMENT GENERATION THROUGH LOANS (NUMBER OF JOBS)			AVERAGE LOAN AMOUNT DISBURSED (AF)		
	Female	Male	Total	Female	Male	Total
Agriculture	310	697	1,007	5,828	7,589	7,047
Beekeeping	12	10	22	5,167	5,500	5,318
Carpentry	13	64	77	6,797	10,789	10,115
Dairy products	37	10	47	4,108	8,520	5,047
Emergency	922	422	1,344	3,790	5,242	4,246
Fishing	1	2	3	10,000	13,000	12,000
Handicrafts	1,945	75	2,020	4,458	7,780	4,582
Heavy machinery	8	5	13	8,563	26,400	15,423
Livestock	7,102	5,169	12,271	6,584	9,739	7,913
Poultry	1,283	154	1,437	3,294	7,397	3,734
Shopkeeping	2,055	2,608	4,663	5,953	8,811	7,552
Small business	2,334	1,936	4,270	4,702	8,196	6,286
Small machinery	1,344	149	1,493	4,280	8,087	4,660
Total	17,366	11,301	28,668	5,430	8,902	7,225

Source: Project documents 2016.

TABLE 4.5 **Employment generation through enterprise group by sector**

SECTOR	DIRECT EMPLOYMENT (NUMBER OF JOBS)			INDIRECT EMPLOYMENT (NUMBER OF JOBS)			AVG. MONTHLY REVENUE (AF)		
	Female	Male	Total	Female	Male	Total	Female	Male	Total
Agriculture	95	285	380	19	0	19	56,614	25,328	32,729
Aluminum	3	4	7	5	0	5	1,691	70,800	56,978
Aquaculture	6	7	13	4	0	4	400	13,417	8,210
Automobile	3	38	41	0	0	0	0	3,860	3,378
Aviculture	637	368	1,005	299	11	310	14,268	43,153	25,549
Business services	25	84	109	18	0	18	2,369	25,779	20,181
Carpentry	0	121	121	0	2	2	0	29,576	29,576
Chemical	14	7	21	0	0	0	27	4,958	1,999
Construction materials	0	31	31	0	0	0	0	36,013	36,013
Cosmetics	45	6	51	35	0	35	12,471	33,538	17,152
Dairy	562	265	827	290	64	354	8,491	7,883	8,313
Electronics	0	11	11	0	0	0	0	8,160	8,160
Energy production	12	0	12	4	0	4	1,077	0	1,077
Food industries	156	167	323	111	3	114	7,866	168,541	102,381
Handicrafts	1,910	217	2,127	1,008	22	1,030	15,206	23,098	16,168
Iron and metallic	0	37	37	0	0	0	0	21,257	21,257
Manufacturing	1	6	7	6	0	6	5,000	92,220	48,610
Packaging	0	13	13	0	0	0	0	12,157	12,157
Recycling	5	0	5	8	0	8	5,000	0	5,000
Total	3,476	1,673	5,149	1,807	102	1,909	14,229	46,589	26,227

Source: Project documents 2016.

AREDP also provides technical and financial support to local SMEs (tables 4.6 and 4A.7). Our research of this support revealed the following:

- About half of the jobs created through SMEs are seasonal, and most are created in Bamyan province. Most jobs are for men, implying lower participation of rural women in SMEs;
- In general, it requires less capital to generate jobs for women than men.

As shown in table 4.7 female entrepreneurs had faster income growth than male entrepreneurs. Income growth for women was very high in Bamyan province, but much lower in Herat (table 4A.7).

Although AREDP is one of the most successful programs in Afghanistan in terms of achieving its employment creation goals and targets, it covers only about 12 percent of the total rural population. Many challenges constrain entrepreneurial activities, particularly in rural areas, including weak marketing infrastructure, low capacities in business service provision, weak agricultural value chains, and lack of access to finance. In this challenging environment, AREDP is successfully providing financial and technical support to rural people to improve enterprises and reduce unemployment and vulnerability. It has huge potential to promote rural nonfarm activities by scaling up its coverage, and also has potential to connect rural women to markets via financial support and training programs.

TABLE 4.6 **Small and medium-sized enterprise: direct and indirect employment, by gender and sector**

Number of jobs

GENDER AND SECTOR	DIRECT	INDIRECT	SEASONAL	TOTAL	AVERAGE CAPITAL (AF) PER JOB
Female	**807**	**555**	**573**	**1,935**	**73,565**
Agriculture	79	28	52	159	109,160
Food	46	9	20	75	59,967
Food processing	60	19	10	89	6,049
Handicrafts	622	500	491	1,613	65,070
Male	**2,719**	**5,524**	**9,817**	**18,060**	**329,268**
Agriculture	579	1,925	844	3,348	413,453
Aluminum	70	170	30	270	45,926
Aquaculture	15	1	16	32	1,860,114
Aviculture	157	523	101	781	340,175
Dairy	141	1,091	34	1,266	114,096
Energy	51	434	23	508	31,439
Food	124	11	2	137	473,472
Food processing	230	467	83	780	221,763
Handicrafts	889	535	8,420	9,844	42,113
Leather	42	39	192	273	15,793
Manufacturing	34	2	20	56	381,764
Marble	173	206	0	379	386,491
Mining	20	0	0	20	510,000
Packaging	30	0	20	50	210,650
Textile and fabric	109	9	0	118	125,489
Tissue paper	50	110	0	160	31,250
Wool	5	0	32	37	9,349
Total	**3,526**	**6,079**	**10,390**	**19,995**	**266,488**

Source: Project documents 2016.

TABLE 4.7 **Average income growth, July 31, 2015–March 31, 2016**

SECTOR	FEMALE (%)	MALE (%)	TOTAL (%)
Agribusiness/agriculture	20.7	52.9	52.1
Apiculture (honey)	34.6	72.8	68.4
Aviculture (poultry)	—	98.8	98.8
Carpet	—	77.8	77.8
Dairy	15.6	176.0	164.6
Floriculture (nursery)	—	38.3	38.3
Food processing	80.4	22.4	40.5
Handicrafts	21.4	37.4	25.4
Livestock	—	29.3	29.3
Manufacturing	—	23.1	23.1
Textile and fabric	—	52.5	52.5
Total	132.3	65.6	78.4

Source: Project documents 2016.
Note: — = not available.

Rural livelihoods and community-driven development: the role of National Solidarity Program III

NSP III was designed to boost rural livelihoods in Afghanistan with the CDD approach, a powerful tool for sustainable and inclusive rural development. NSP III built gender-balanced, community-based institutions, supporting them in the design and implementation of community subprojects. These subprojects generated direct short-term employment and livelihood opportunities for rural people and set the stage to help local economies flourish in the long run. The NSP III project ended in 2017 and its successor, Citizens Charter, has an even broader agenda for inclusive rural development. The new program will generate long-term sustainable jobs and livelihood opportunities.

Background. The MRRD initiated NSP in 2003 with support from the World Bank and other donor agencies to empower Afghanistan's rural communities to identify, plan, and manage their own development projects. NSP was designed to build representative gender-balanced institutions for local governance and deliver critical services to the rural population. It built CDCs, provided them block grants to fund subprojects, and linked them with government agencies and NGOs to improve access to services and resources. The coverage area included all of Afghanistan's rural communities. NSP III, which closed in March 2017, was expected to generate more than 29.2 million labor days, or 146,092 FTE jobs. In total, there were more than 1.2 million direct beneficiaries, of whom about 46 percent were female. CDCs are operating or being established in 373 of 408 districts and provincial centers in all 34 provinces.

Maintenance Cash Grant Project: Jobs for Peace Initiative. The MCG Project launched in 2016 as a part of the Afghan Government's Jobs for Peace Initiative, with the primary objective of creating short-term employment for the neediest households in rural communities. A secondary objective was to address communities' maintenance problems through grants for essential infrastructure repairs. To date, NSP III has completed more than 72,000 infrastructure projects. The MCG pilot program, initially funded with $50 million, was undertaken to ensure maintenance of this infrastructure, and is expected to operate in roughly 4,700 communities. The maintenance grants are labor-intensive projects with approximately 70 percent of funds dedicated to labor and 30 percent to material inputs. Thus, the MCG has become an important instrument for job creation in rural areas. It is implemented in 12 provinces—Baghlan, Balkh, Farah, Faryab, Herat, Jawzjan, Kandahar, Khost, Kunarha, Kunduz, Laghman, and Nangarhar—selected based on the extent of underemployment, seasonality, access during the winter months, and security.

The overall NSP III goal of was to promote socioeconomic development in rural communities. Its other goals were to expand to the 10,320 communities it did not initially cover and to scale up support for local economic and social development through the provision of repeater block grants to 17,400 rural communities.

Key inputs and outcomes. According to the 2017 ISR report, NSP III exceeded many of its goals, as follows:

• The goal was that at least 70 percent of sampled communities would recognize CDCs as legitimate institutions, representative in decision-making and in the development of their communities. Currently, 96 percent (13,428) of sampled communities recognize this.

- About 78 percent (2,683) of CDCs performed their functional mandates, such as conducting meetings, sharing information, building linkages, and successfully completing subprojects. The goal was 60 percent.
- The target was for at least 70 percent of sampled communities to have improved access to services (irrigation, power, transport, water supply, and other facilities). By December 31, 2015, about 80 percent of the sampled communities had access to such services.
- The goal was to ensure that at least half of beneficiaries were women. The figure was 48 percent in December 31, 2016.
- As of December 31, 2016, interventions had built or rehabilitated 2,995 additional classrooms at the primary level.
- By December 31, 2016, almost 12,500 kilometers of rural roads had been constructed; 141 kilometers had been rehabilitated.
- More than 5 million people in rural areas were provided with improved access to water, and more than 6,000 improved latrines were constructed.
- More than 361,000 hectares of land had been brought under irrigation and drainage services by December 31, 2016.

Job creation effects. NSP III is run by 987 staff (89 percent male, 11 percent female) (table 4A.5). Seventy-eight percent are skilled workers and 22 percent are unskilled workers such as guards and cleaners. The program generated about 137,000 FTE jobs directly and 32,000 indirect jobs (table 4.8). Most direct jobs were generated in the transport, irrigation, and rural development sectors. Many subprojects focused on transport, highlighting that sector's importance for connecting rural people with markets. Indirect jobs were created mostly due to the activities in the rural development and water supply sectors. Most subprojects in irrigation, education, power, and transport were prone to create more induced employment, as they were expected to substantially increase local productive capacity. The MCG pilot program created about 22,400 FTE jobs, mostly in the irrigation and transportation sectors. Figure 4A.6 in annex 4A shows that NSP III did not create many jobs in some provinces with high unemployment rates (such as Bamyan and Daykundi). Job creation was highest in Panjsher province, followed by Khost, Herat, and Kunarha.

TABLE 4.8 National Solidarity Program III job creation

SECTOR	AVERAGE SUBPROJECT BUDGET (US$)	NO. OF BENEFICIARY FAMILIES	NO. OF SUBPROJECTS	DIRECT FTE JOBS	INDIRECT FTE JOBS[a]	MCG PROJECT FTE JOBS (JOBS FOR PEACE)		TOTAL FTE JOBS
						Skilled	Unskilled	
Education	27,851	212,524	785	4,935	694	17	128	5,774
Health	34,836	23,355	95	556	89	1	5	651
Irrigation	22,588	1,830,217	9,104	33,972	847	147	4,036	39,002
Livelihood	10,307	19,950	101	190	99	—	—	289
Power	23,395	360,843	1,644	8,501	1,093	13	179	9,786
Rural dev.	18,483	830,676	3,981	16,363	3,937	41	390	20,731
Transport	26,559	2,300,220	10,356	44,681	—	332	16,037	61,250
Water supply & sanitation	23,201	1,863,444	8,676	28,044	2,402	40	833	31,319
Total	23,402	7,441,229	34,742	137,241	9,161	591	21,808	168,801

Source: Project documents 2016.
Note: FTE = full-time equivalent; MCG = Maintenance Cash Grant; — = not available.
a. Estimated by assuming one subproject equaled one FTE job for the subprojects with job creation potential.

Spending was high for subprojects in the health, power, rural development, and transport sectors. Spending was lowest in the agriculture sector, followed by the livelihoods program. NSP III spending per beneficiary was also low in sub-projects related to agriculture and livelihoods (see annex 4A for details of NSP III beneficiaries and job creation). Results indicate that NSP III generated many direct jobs, mostly for unskilled rural workers, most of which were unsustainable. Indirect job creation was lower. The MCG Pilot program had a significant job creation effect, but its coverage was limited.

JOBS FROM PRIVATE SECTOR INTERVENTIONS: EVIDENCE FROM OMAID BAHAR FRUIT PROCESSING LIMITED

This section provides an initial assessment of the architecture and number of jobs in a selected fruits agro-processing supply chain. Using a case study approach, it explores where jobs exist in the upstream network of suppliers and the downstream network of transporters, distributors, and wholesalers. We have estimated the number of permanent and temporary jobs in each segment of the supply chain. We profile the current workforce occupying these jobs, with information on the type of employment contracts, tenure of engagement, training opportunities, average daily wages, demographic distribution, educational qualifications, and working conditions.

Omaid Bahar Fruit Processing Limited, located in Kabul, was chosen for the case study for two reasons. First, it has a significant domestic market share of the fruit processing sector, accounting for more than Af 892 million ($13.25 million) in revenues and hundreds of formal jobs. Second, it met the criteria that the lead firm must have benefitted from a private sector support operation financed by the International Finance Corporation between 2012 and 2014. This development relationship facilitated collaboration with Omaid Bahar's chief executive and management, considered necessary for outreach and data collection from upstream and downstream linkage firms.

Background. The study is based on primary data collection from a proportional sample of entities, including individuals and SMEs that constitute Omaid Bahar's upstream and downstream linkages. The case study's activities can be grouped into four stages: (a) Profile upstream and downstream linkages (November 2016); (b) Survey design, instruments, sampling (December 2016–January 2017); (c) Face-to-face data collection (February 2017); and, (d) Analysis and reporting (April 2017)

The fruit processing supply chain in Afghanistan was profiled to include five categories of participants: suppliers, transporters, processors, distributors, and wholesalers. Early consultations with Omaid Bahar were used to identify its upstream and downstream linkages in the fruit processing sector, which were also consistent with industry linkages in general. The entities identified in this first stage can be viewed as constituent parts in a traditional supply chain. Supply chains usually begin with suppliers, involve intermediaries such as manufacturers, processors, distributors, transporters, wholesalers, and retailers, and conclude at the paying customer. In addition, we profiled Omaid Bahar using targeted questions around revenues from sales, sources of inputs and supplies, and workforce composition and profile.

Sampling methodology. The study employed a proportional sampling approach covering 31 firms across sections of the fruit processing supply chain.

This approach was informed by a variety of factors, including a tight timeline, a limited budget for data collection, and logistical challenges in remote areas. For suppliers and transportation segments of the supply chain, a random sample was drawn of 33 percent of the firms. The list of wholesalers, supplied by Omaid Bahar's distributors, included 120 firms. Interviewing a sample of 33 percent would have meant interviewing 40 wholesalers, including some in fragile and remote areas, which would not have been practical. Therefore, considering the homogeneity of economic structures in each province, the team designed a sample of distributors and wholesalers based on qualitative criteria, not a random selection. To do this, distributors that cover multiple provinces were selected to represent each type of "provincial profile."

Three linked wholesalers were randomly identified for each distributor, resulting in 15 wholesalers. Given Omaid Bahar's market share, there were likely hundreds of formal and informal retailers carrying its processed fruit products. Furthermore, retail outlets in Afghanistan are usually small, independently owned by families, and employ a handful of people. They stock and sell many products, of which Omaid Bahar processed fruits products were hypothesized to account for a small share (single-digit percentages) of their sales, revenues, and profit margins. This has implications for our study and jobs estimation.

Consequently, fewer jobs were supported in the retailer section of the supply chain than in other parts, where business and revenue share attributed to Omaid Bahar are significantly higher. During consultations with management, the study team learned that product sales were concentrated in urban areas. Operating within the time and budget constraints limited field data collection; as a result, a convenience sample of 20 retailers covering major localities in the Kabul urban area was interviewed with a smaller set of questions. Questions included an estimate of monthly revenues, share of sales revenue attributed to Omaid Bahar processed fruit products, the share of average margin on these products, and the total number of employees in the retail establishment.

The selection of distributors was based on the coverage of each provincial profile, namely urban center and rural province, historical breadbasket and agriculture-intensive, and dynamic economies. In addition, sample selection was based on the following criteria: (a) Geographical coverage of most of the country; (b) A balanced number of urban centers and rural areas; (c) A balanced number of economically dynamic provinces and weakened economies; (d) A balanced number of historically agricultural provinces and breadbaskets versus other provinces; and, (e) Covering distributors for five provinces—Herat, Jalalabad, Kabul, Kandahar, and Mazar-e-Sharif—allowed the study team to accommodate all the above criteria (table 4.9).

The selection of the five distributors in major urban centers provided geographical coverage for most of the country, including both large urban centers and more rural areas. These companies are based in all the major urban centers and ensure the distribution of Omaid Bahar's products in rural areas (for example, Helmand, Jozjan, and Laghman). Primary data at the firm-level were collected to support estimation of jobs in sections of Omaid Bahar's fruit processing supply chain. Data were collected using a face-to-face survey.

Estimation and architecture of jobs in the fruit processing supply chain. Using case study results, we estimated the number of jobs supported in firms' supply chain networks, which form a continuum of resources and materials that

TABLE 4.9 **Selection of distributors based on provincial profiles**

DISTRIBUTION AREAS	URBAN CENTER AND RURAL PROVINCE	HISTORICAL BREADBASKET	DYNAMIC ECONOMY
Jalalabad and Laghman	✓	✓	✓
Kandahar and Helmand	✓	✓	X
Herat, Farah, and Baghdis	✓	X	✓
Mazar e Sharif, Jozjan, and Samangan	✓	X	X
Kabul	X	✓	✓
Ghazni and Zaboul	X	X	X
Noristan and Laghman	X	✓	X
South	X	X	X
Puli Khumri and Kunduz	X	X	X

Source: Based on the report's sample survey, 2017.

TABLE 4.10 **Permanent and temporary jobs in the fruit-processing supply chain, five-year average**

SUPPLY CHAIN SEGMENT	PERMANENT	PERCENT	TEMPORARY	PERCENT	ROW TOTAL	PERCENT
Input supplier	25	6.2	57	11.8	82	9.3
Service provider	3	0.7	0	0.0	3	0.3
Transporter	32	8.0	10	2.0	42	4.7
Lead firm	137	34.3	342	71.0	479	54.4
Distributor	106	26.6	43	9.0	149	17.0
Wholesaler	96	24.1	30	6.2	126	14.3
Total	399	100.0	482	100.0	881	100.0

Source: Based on the report's sample survey, 2017.

flow progressively from the origin of its constituent materials to the customer. Each section of this network plays a role in transmitting materials, intermediary products and services, and the final goods, satisfying market demand downstream and leveraging margin opportunities. Each section, viewed as upstream or downstream from relative positions around the lead firm, provides economic opportunities for a variety of workers. The network supports permanent and temporary jobs that offer opportunities for workers to engage in income-generating activities in exchange for labor.

We found that over the past five fiscal years (2012–16), the upstream and downstream network supported 881 jobs (approximately 8.5 jobs for every 10 jobs at Omaid Bahar), more than half of which supported the firm's fruit processing activities. Of the remaining jobs, most were with distributor firms (17 percent), followed by wholesalers (14 percent) and suppliers (9 percent). These aggregate totals do not reflect the heterogeneity in permanent and seasonal jobs. Table 4.10 presents information differentiated by the nature of employment contract (for example, permanent or temporary) and lists averages over the five-year period.

In the most recent fiscal year (2016), there were 983 total jobs in the supply chain, about a 36 percent increase over the past five years. This growth is mainly due to workforce expansion at Omaid Bahar (from 340 to 600). Since fiscal year 2012, the firm has expanded its workforce by about 76 percent, nearly doubling the number of permanent contracts (from 90 to 175) and realizing a substantial increase in temporary contracts (from 250 to 425).

ANNEX 4A

TABLE 4A.1 **Donor community investments in agriculture in Afghanistan 2000–2016**

DONORS	PROJECT BUDGET (US$)			PROJECT COUNT			REGION									
	Active	Closed	Grand total	Active	Closed	Grand total	North-east***	West central***	East***	South**	North**	West**	Southwest*	Central*	National	Unspecified
ADB	**165,500,000**	**59,875,000**	**225,375,000**	**5**	**3**	**8**	**5**	**2**	**3**	—	**5**	**2**	**2**	**4**	**2**	**1**
Agriculture production & marketing	29,500,000	58,000,000	87,500,000	2	2	4	2	2	3	—	4	2	2	4	—	—
Irrigation	136,000,000	1,875,000	137,875,000	3	1	4	3	—	—	—	1	—	—	—	2	1
Australia	**42,000,000**	**26,000,000**	**68,000,000**	**2**	**2**	**4**	**3**	**2**	—	—	**3**	**1**	**1**	**1**	**2**	—
Agriculture production & marketing	17,000,000	26,000,000	43,000,000	1	2	3	—	—	—	—	—	—	1	—	2	—
CDD	25,000,000		25,000,000	1	—	1	1	1	—	—	1	1	—	1	—	—
Canada	—	**75,000,000**	**75,000,000**	—	**1**	**1**	—	—	—	—	—	—	—	—	**1**	—
Agriculture production & marketing	—	75,000,000	75,000,000	—	1	1	—	—	—	—	—	—	—	—	1	—
CERP project	—	**352,105**	**352,105**	—	**1**	**1**	—	—	—	—	—	—	—	**1**	—	—
Agriculture production & marketing	—	352,105	352,105	—	1	1	—	—	—	—	—	—	—	1	—	—
EU	**6,900,000**	**77,553,815**	**84,453,815**	**1**	**8**	**9**	**6**	**3**	**5**	**3**	**7**	**3**	**3**	**6**	**1**	**3**
Agriculture production & marketing	6,900,000	27,307,370	34,207,370	1	5	6	5	2	5	3	7	3	3	6	—	3
Capacity building	—	7,000,000	7,000,000	—	1	1	—	—	—	—	—	—	—	—	1	—
Irrigation	—	43,246,445	43,246,445	—	2	2	1	1	—	—	—	—	—	—	—	—
FAO	**12,300,000**	**29,161,614**	**41,461,614**	**1**	**9**	**10**	**3**	**1**	**2**	—	—	—	—	**1**	**4**	**1**
Agriculture production & marketing	—	25,059,064	25,059,064	—	7	7	2	—	2	—	—	—	—	1	2	1

continued

TABLE 4A.1, continued

DONORS	PROJECT BUDGET (US$)			PROJECT COUNT			REGION									
	Active	Closed	Grand total	Active	Closed	Grand total	North-east***	West central***	East****	South**	North**	West**	Southwest*	Central*	National	Unspecified
Capacity building	—	3,452,550	3,452,550	—	1	1	—	—	—	—	—	—	—	—	1	—
CDD	—	—	—	—	—	—	—	—	—	—	—	—	—	—	—	—
Food security	—	650,000	650,000	—	1	1	—	—	—	—	—	—	—	—	1	—
Irrigation	12,300,000	—	12,300,000	1	—	1	—	1	—	—	—	—	—	—	—	—
France	**9,424,852**	**10,920,000**	**20,344,852**	**2**	**2**	**4**	—	—	—	—	—	—	—	**2**	—	**3**
Agriculture production & marketing	9,424,852	10,920,000	20,344,852	2	2	4	—	—	—	—	—	—	—	2	—	3
Germany	**8,050,000**	**4,934,563**	**12,984,563**	**1**	**2**	**3**	2	—	1	—	1	—	—	—	1	—
Agriculture production & marketing	8,050,000	2,677,546	10,727,546	1	1	2	2	—	1	—	1	—	—	—	—	—
Capacity building	—	2,257,017	2,257,017	—	1	1	—	—	—	—	—	—	—	—	1	—
IFAD	**32,100,000**	**5,088,000**	**37,188,000**	**1**	**2**	**3**	2	1	2	—	2	—	—	1	—	—
Agriculture production & marketing	32,100,000	5,088,000	37,188,000	1	2	3	2	1	2	—	2	—	—	1	—	—
Italy	**5,750,000**	**4,000,000**	**9,750,000**	**1**	**1**	**2**	—	1	2	—	—	2	—	1	—	—
Agriculture production & marketing	5,750,000	4,000,000	9,750,000	1	1	2	—	1	2	—	—	2	—	—	—	—
Japan	**2,000,000**	**24,100,000**	**26,100,000**	**1**	**3**	**4**	1	1	2	—	2	—	—	1	—	1
Agriculture production & marketing	2,000,000	24,100,000	26,100,000	1	3	4	1	1	2	—	2	—	—	1	—	1
MDG achievement fund	**5,000,000**	—	**5,000,000**	**1**	—	**1**	1	1	1	—	—	—	—	1	—	—

continued

TABLE 4A.1, *continued*

DONORS	PROJECT BUDGET (US$)			PROJECT COUNT			REGION									
	Active	Closed	Grand total	Active	Closed	Grand total	North-east***	West central***	East***	South**	North**	West**	Southwest*	Central*	National	Unspecified
Food security	5,000,000	—	5,000,000	1	—	1	1	1	1	—	—	—	—	1	—	—
Norway	—	**7,800,000**	**7,800,000**	—	**1**	**1**	—	—	—	—	—	—	—	—	—	**1**
Agriculture production & marketing	—	7,800,000	7,800,000	—	1	1	—	—	—	—	—	—	—	—	—	1
Spain	—	**2,500,194**	**2,500,194**	—	**1**	**1**	—	**1**	—	—	—	**1**	—	—	—	—
Environment management	—	2,500,194	2,500,194	—	1	1	—	1	—	—	—	1	—	—	—	—
UK DFID	**36,456,750**	**95,465,531**	**1 31,922,281**	**1**	**10**	**11**	—	**3**	—	—	—	—	**5**	—	**4**	—
Agriculture production & marketing	36,456,750	95,465,531	1 31,922,281	1	10	11	—	3	—	—	—	—	5	—	4	—
UNDP	—	**19,999,406**	**19,999,406**	—	**1**	**1**	**1**	—	**1**	—	**1**	**1**	**1**	—	—	—
CDD	—	19,999,406	19,999,406	—	1	1	1	—	1	—	1	1	1	—	—	—
USAID	**443,481,033**	**353,000,000**	**796,481,033**	**9**	**3**	**12**	**1**	**3**	**5**	—	**2**	**5**	**7**	**1 2**	**3**	**2**
Agriculture production & marketing	384,331,033	353,000,000	737,331,033	7	3	10	—	1	2	—	2	4	7	8	2	2
Capacity building	19,900,000	—	19,900,000	1	—	1	—	—	—	—	—	—	—	—	1	—
Irrigation	39,250,000	—	39,250,000	1	—	1	1	—	1	—	—	1	—	4	—	—
Grand total	**768,962,635**	**795,750,228**	**1,564,712,862**	**26**	**50**	**76**	**25**	**19**	**22**	**3**	**23**	**15**	**19**	**30**	**18**	**12**

Source: Project documents of donor entities from respective institutional portals, donors' consultation meeting, Kabul, 2016.

Note: Classification of provinces by regions is as follows: Southwest:: Nimroz, Helmand, Kandahar, Zabul, Urozgan; Central: Kabul, Kapisa, Parwan, Wardak, Logar, Panjsher; West: Badghis, Herat, Farah; North: Samangan, Balkh, Jawzjan, Sar-e-Pul, Faryab; South: Ghazni, Paktika, Paktya, Khost; East: Nangarhar, Kunarha, Laghman, Nooristan; West central: Ghor, Bamyan, Daykundi; Northeast: Badakhshan, Takhar, Baghlan, Kunduz. Poverty headcount (%) by region is as follows: *(0–30), **(30–40), ***(40–50).

TABLE 4A.2 Donor community investments in agriculture 2000–2016, per type of project intervention

PROJECT INTERVEN-TIONS	PROJECT BUDGET (US$)			PROJECT COUNT			PROVINCE COUNT PER REGION									
	Active	Closed	Grand total	Active	Closed	Grand total	Northeast***	West central***	East***	South**	North**	West**	Southwest**	Central*	National	Unspecified
Agriculture production & marketing	531,512,635	714,769,616	1,246,282,251	18	41	59	16	11	17	3	20	13	18	24	11	11
ADB	29,500,000	58,000,000	87,500,000	2	2	4	2	2	3	–	4	2	2	4	–	–
Australia	17,000,000	26,000,000	43,000,000	1	2	3	–	–	–	–	–	–	1	–	2	–
Canada	–	75,000,000	75,000,000	–	1	1	–	–	–	–	–	–	–	–	1	–
CERP project	–	352,105	3 52,105	–	1	1	–	–	–	–	–	–	–	1	–	–
EU	6,900,000	27,307,370	34,207,370	1	5	6	5	2	5	3	7	3	3	6	–	3
FAO	–	25,059,064	25,059,064	–	7	7	2	–	2	–	–	–	–	1	2	1
France	9,424,852	10,920,000	20,344,852	2	2	4	–	–	–	–	–	–	–	2	–	3
Germany	8,050,000	2,677,546	10,727,546	1	1	2	2	–	1	–	1	–	–	–	–	–
IFAD	32,100,000	5,088,000	37,188,000	1	2	3	4	1	2	–	4	–	–	1	–	–
Italy	5,750,000	4,000,000	9,750,000	1	1	2	–	1	–	–	–	4	–	–	–	–
Japan	2,000,000	24,100,000	26,100,000	1	3	4	1	1	2	–	2	–	–	1	–	1
Norway	–	7,800,000	7,800,000	–	1	1	–	–	–	–	–	–	–	–	–	1
UK DFID	36,456,750	95,465,531	131,922,281	1	10	11	–	3	–	–	–	–	5	–	4	–
USAID	384,331,033	353,000,000	737,331,033	7	3	10	–	1	2	–	2	4	7	8	2	2
Capacity building	19,900,000	12,709,567	32,609,567	1	3	4	–	–	–	–	–	–	–	–	**4**	–
EU	–	7,000,000	7,000,000	–	1	1	–	–	–	–	–	–	–	–	1	–
FAO	–	3,452,550	3,452,550	–	1	1	–	–	–	–	–	–	–	–	1	–
Germany	–	2,257,017	2,257,017	–	1	1	–	–	–	–	–	–	–	–	1	–
USAID	19,900,000	–	19,900,000	1	–	1	–	–	–	–	–	–	–	–	1	–
CDD	25,000,000	19,999,406	44,999,406	1	1	2	2	1	1	–	2	2	1	1	–	2
Australia	25,000,000	–	25,000,000	1	–	1	1	1	–	–	1	1	–	–	–	–
FAO	–	–	–	–	–	–	–	–	–	–	–	–	–	–	–	–
UNDP	–	19,999,406	19,999,406	–	1	1	1	–	1	–	1	1	1	–	–	2

continued

TABLE 4A.2, *(continued)*

PROJECT INTERVEN-TIONS	PROJECT BUDGET (US$)			PROJECT COUNT			PROVINCE COUNT PER REGION									
	Active	Closed	Grand total	Active	Closed	Grand total	Northeast***	West central***	East***	South**	North**	West**	Southwest*	Central*	National	Unspecified
Environment management	—	2,500,194	2,500,194	—	1	1	—	1	—	—	—	1	—	—	—	—
Spain	—	2,500,194	2,500,194	—	1	1	—	1	—	—	—	1	—	—	—	—
Food security	5,000,000	650,000	5,650,000	1	1	2	1	1	1	—	—	—	—	1	1	—
FAO	—	650,000	650,000	—	1	1	—	—	—	—	—	—	—	—	1	—
MDG achievement fund	5,000,000	—	5,000,000	1	—	1	1	2	1	—	—	—	—	1	—	—
Irrigation	187,550,000	45,121,445	232,671,445	5	3	8	4	3	1	—	1	1	—	1	2	1
ADB	136,000,000	1,875,000	137,875,000	3	1	4	1	—	—	—	1	—	—	—	2	1
EU	—	43,246,445	43,246,445	—	2	2	1	1	—	—	—	—	—	—	—	—
FAO	12,300,000	—	12,300,000	1	—	1	1	1	1	—	—	—	—	—	—	—
USAID	39,250,000	—	39,250,000	1	—	1	1	1	1	—	—	1	—	1	—	—
Grand total	768,962,635	795,750,228	1,564,712,862	26	50	76	23	17	20	3	23	17	19	27	18	12

Source: Project documents of donor entities from respective institutional portals, donors' consultation meeting, Kabul, 2016.

Note: Classification of provinces by regions is as follows: Southwest: Nimroz, Helmand, Kandahar, Zabul, Urozgan; Central: Kabul, Kapisa, Parwan, Wardak, Logar, Panjsher; West: Badghis, Herat, Farah; North: Samangan, Balkh, Jawzjan, Sar-e-Pul, Faryab; South: Ghazni, Paktika, Paktya, Khost; East: Nangarhar, Kunarha, Laghman, Nooristan; West central: Ghor, Bamyan, Daykundi; Northeast: Badakhshan, Takhar, Baghlan, Kunduz. Poverty headcount (%) by region is as follows: *(0–30), **(30–40), ***(40–50). — = not available.

TABLE 4A.3 **Indicators to improve job monitoring**

COUNTRY	IMPACT STATEMENT	INDICATIVE IMPACT INDICATORS
Afghanistan	Sustainable improvements in the livelihoods of people through reduced vulnerability to shocks, increased incomes, and access to sustainable and inclusive jobs	1. Number of jobs (disaggregated by gender, age, socioeconomic status) supported by World Bank interventions 2. Percentage increase in incomes supported by World Bank interventions (disaggregated by gender, age, socioeconomic status) 3. Improved working conditions supported by World Bank interventions (percent of workers covered by social benefits)

INTERVENTION TYPE	OUTCOME STATEMENT	INDICATIVE OUTCOME INDICATORS
Agricultural production and marketing interventions	Improved agricultural technique and product quality; improved farmer-market linkages	1. Percentage of farmers with access to inputs, services, market incentives 2. Percentage of farmers with access to regional/provincial markets
Irrigation and on-farm water management	Increased yield supported by improvement of irrigation system management and diversification to higher-value crops	1. Percentage increase of land under irrigation (arable land) 2. Number of farmers adopting efficient farm management practices 3. Percentage change in annual expenditures on operation and maintenance 4. Production volume of high-value crops 5. Number of operational water user associations created
Rural enterprises	Strengthened forward and backward linkages to rural economy; diversification to off-farm economy; technology adoption by small and medium-sized enterprises	1. Number of rural entrepreneurs with access to credit/working capital 2. Number of rural entrepreneurs with access to technologies 3. Number of targeted clients who are members of an association (for example, producer association or cooperative)
Rural livelihoods and community-driven development	Improved human and social capital for farmers and rural entrepreneurs; strengthened access to regional/provincial markets	1. Number of farmers/rural entrepreneurs with access to finance 2. Outstanding rural microfinance loan portfolio (dollar amount)/ number of active microfinance loan accounts of holders domiciled in rural areas (disaggregated by gender of holder) 3. Number of farmers/rural entrepreneurs with access to basic services (market/access to roads/ infrastructure), education 4. Number of client days of extension services provided to farmers, community members, et al.

Sources: Global Donor Platform for Rural Development, World Bank, and FAO 2008; World Bank 2013; World Bank project documents; and World Bank Operations Policy and Country Services Corporate Results Indicators.

The National Horticulture and Livestock Project

In most provinces, maximum jobs generated through NHLP horticulture activities are the outcome of direct job creation and are long-term jobs (figure 4A.1). Job creation effects are high in Balkh, Kabul, and Samangam.

Although it is expected that NHLP's livestock extension activities have important job creation effects, we do not have information about the actual number of labor days they have generated. Many farmers benefitted from project support for animal health and extension activities. Figure 4A.2 shows that, in most districts, more than 50,000 farmers benefitted from NHLP's animal health and extension activities. These extension services play an important role in improving the labor productivity of workers in the livestock sector, many of whom are women and youth, and therefore play a crucial role in generating more jobs.

FIGURE 4A.1

Horticulture jobs created by NHLP by province

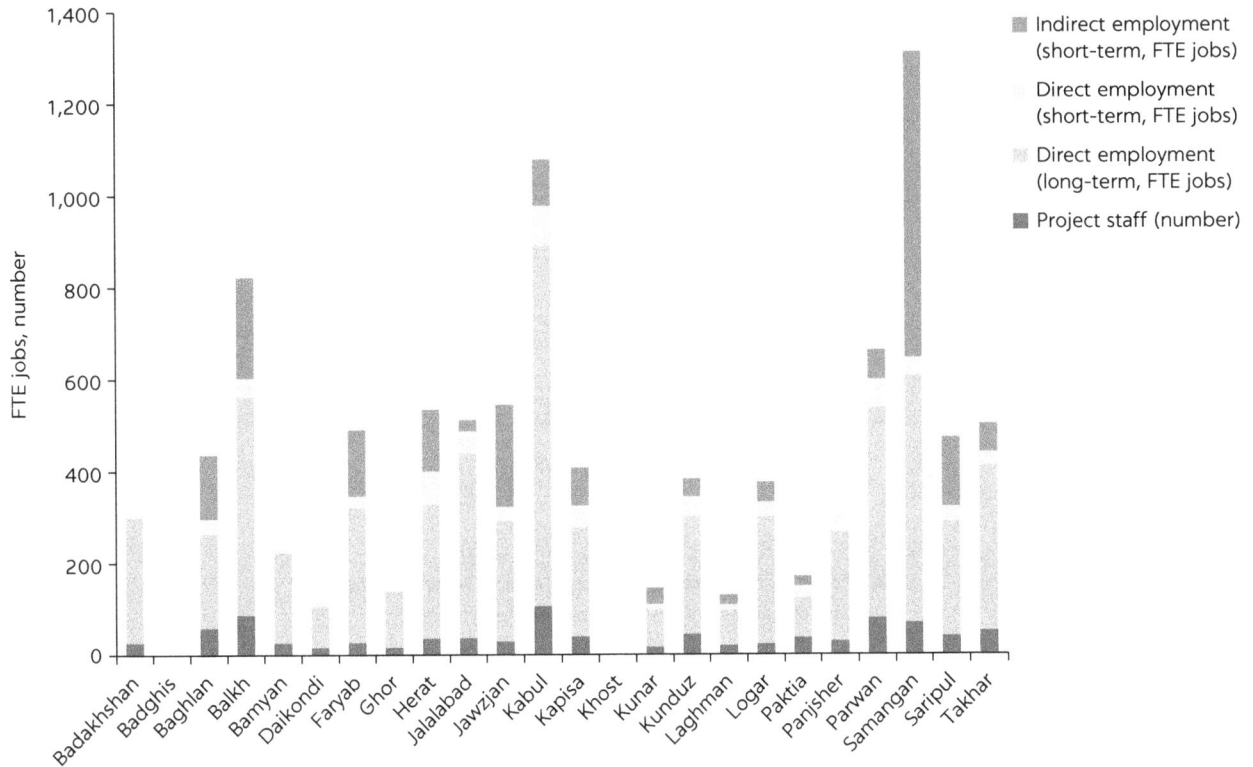

Source: Project documents 2016.

FIGURE 4A.2

Number of farmers benefiting from NHLP animal health and extension activities

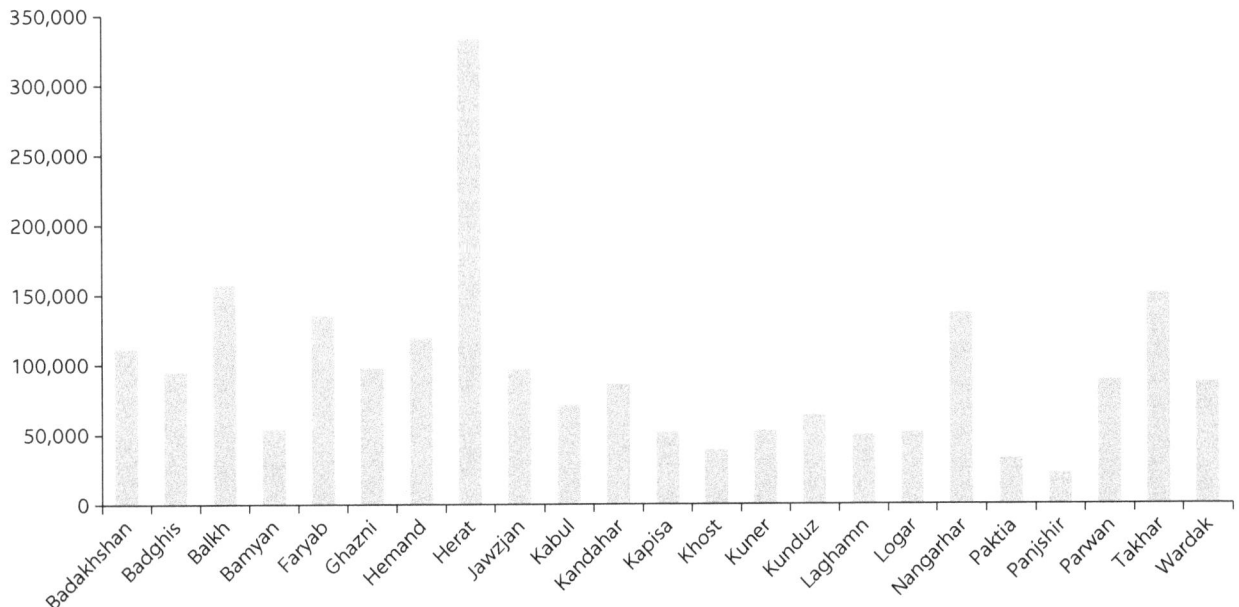

Source: Project documents 2016.

FIGURE 4A.3

Intensity of livestock employment and NHLP beneficiaries, as of Sept. 2016

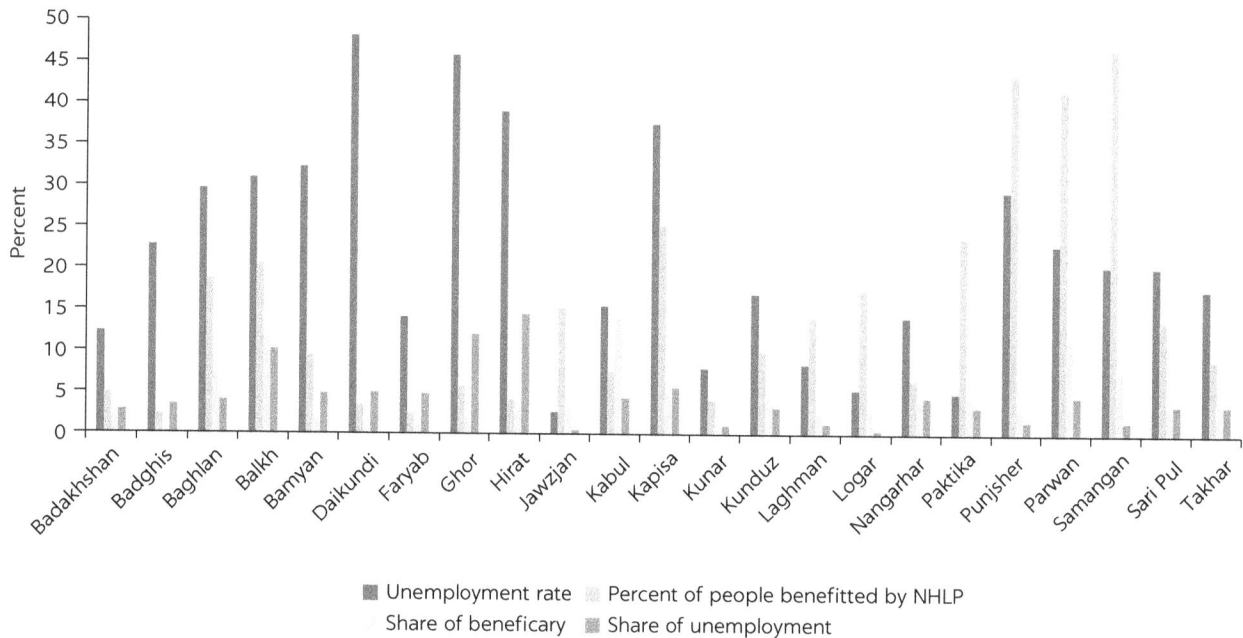

■ Unemployment rate ▨ Percent of people benefitted by NHLP
▨ Share of beneficary ▨ Share of unemployment

Source: Project documents 2016.

The On-Farm Water Management Project (OFWMP)

TABLE 4A.4 **Beneficiaries of OFWMP by province**

PROVINCE	NO. OF BENEFICIARIES	PERCENT WHO BENEFITTED	AMOUNT SPENT (US$)	NO. OF IRRIGATION ASSOCIATIONS ESTABLISHED
Balkh	22,635	5	1,328,097	32
Samangan	13,002	10	641,362	10
Jowzjan	6,810	3	196,147	6
Fareyab	6,295	2	157,069	5
Herat	17,393	3	2,030,830	63
Baghlan	38,967	9	150,147	35
Bameyan	7,370	3	670,193	14
Kabul	11,379	1	1,877,729	20
Kapisa	7,973	4	104,446	10
Parwan	10,121	8	169,743	8
Nangarhar	14,849	5	628,323	30
Laghman	16,311	11	265,387	17
Konarha	4,060	2	618,998	20
Total	177,165	5	8,838,474	270

Source: Project documents 2016.

The Afghanistan Rural Enterprise Development Project

TABLE 4A.5 **Employment generation through loans by province**

SECTOR	PROVINCES					
	Balkh	Bamyan	Herat	Nangarhar	Parwan	Total
Agriculture	280	148	131	405	43	1,007
Beekeeping	7	3	4	2	6	22
Carpentry	6	40	0	29	2	77
Dairy products	2	20	14	7	4	47
Emergency	997	54	98	104	91	1,344
Fishing	1	0	0	1	1	3
Handicrafts	601	1,005	172	215	27	2,020
Heavy machinery	8	1	0	3	1	13
Livestock	2,374	988	2,491	5,061	1,357	12,271
Poultry	330	62	19	776	250	1,437
Shopkeeping	532	407	1,172	1,834	718	4,663
Small business	598	599	1,691	816	566	4,270
Small machinery	976	5	5	495	12	1,493
Total	6,712	3,332	5,797	9,748	3,078	28,668

Source: Project documents 2016.

TABLE 4A.6 **Employment generation through enterprise group by province**

PROVINCE	DIRECT EMPLOYMENT	INDIRECT EMPLOYMENT			SEASONAL EMPLOYMENT	INCOME GROWTH (%), 2015–2016		
		Male	Female	Total		Male	Female	Total
Balkh	849	117	75	192	165	33.84	74.01	47.67
Bamyan	735	88	27	115	606	106.64	81.96	86.31
Herat	894	53	15	68	0	82.13	82.17	82.17
Nangarhar	1,797	75	96	170	1,262	147.09	83.56	93.99
Parwan	874	12	36	47	11	117.84	76.08	94.89
Total	5,149	344	249	593	2,044	65.48	80.62	75.20

Source: Project documents 2016.

TABLE 4A.7 **SME: Direct and indirect employment by province**

PROVINCE	DIRECT	INDIRECT	SEASONAL	TOTAL	AVERAGE CAPITAL (AF) PER JOB
Balkh	**569**	**2,659**	**558**	**3,786**	**334,275.9**
Female	203	45	154	402	81,553.84
Male	366	2,615	404	3,385	364,289
Bamyan	**263**	**733**	**8,394**	**9,390**	**28,157**
Female	102	235	243	580	15,739
Male	161	498	8,151	8,810	37,470
Herat	**862**	**1,928**	**773**	**3,563**	**419,220**
Female	102	270	44	416	23,958
Male	760	1,658	729	3,147	492,757
Nangarhar	**1,304**	**99**	**421**	**1,824**	**192,024**
Female	373	5	128	506	36,508
Male	931	94	293	1,318	204,698
Parwan	**528**	**661**	**244**	**1,433**	**285,679**
Female	27	1	4	32	117,535
Male	501	660	240	1,401	304,361
Grand total	**3,526**	**6,079**	**10,390**	**19,995**	**266,488**

Source: Project documents 2016.

FIGURE 4A.4

Landlessness, unemployment rate, and intensity of AREDP intervention, as of Sept. 2016

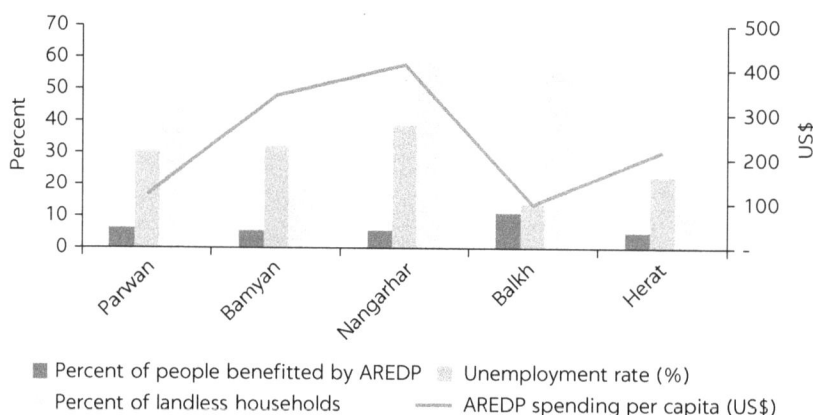

Source: Calculation using Afghanistan Living Condition Survey 2013–14 and project documents 2016.

Figure 4A.4 reveals that fewer than 10 percent of people in most provinces benefitted from AREDP activities and that project coverage does not reflect the severity of unemployment. In Balkh, the unemployment rate is the lowest among the provinces where AREDP operates, yet coverage intensity is highest. However, per capita project spending is highest in Nangarhar province, where both unemployment and landlessness are the highest.

The National Solidarity Program III

Figure 4A.5 shows that the spending per subproject is high in the health, power, rural development, and transport sectors. The subproject cost is lowest in the agriculture sector, followed by the livelihoods program; spending per beneficiary is also low in these subprojects.

We also explored whether there is any association between program spending and employment indicators. Figure 4A.6 explores if provinces severely affected by unemployment were targeted by programs. It shows that unemployment rates are disproportionately high in many provinces, but these provinces could not get adequate project coverage. For example,

TABLE 4A.8 **Direct employment–project staffs**

GENDER	FREQUENCY	PERCENT	
Female	112	11.35	
Male	875	88.65	
Total	987	100	

SKILL	FREQUENCY	PERCENT	
Unskilled	213	21.58	
Skilled	774	78.42	
Total	987	100	

GENDER	UNSKILLED	SKILLED	TOTAL
Female	23	89	112
Percent	20.54	79.46	100
Male	190	685	875
Percent	21.71	78.29	100
Total	213	774	987
Percent	21.58	78.42	100

Source: Project documents 2016.

FIGURE 4A.5

Intensity of NSP spending among beneficiaries, as of Sept. 2016

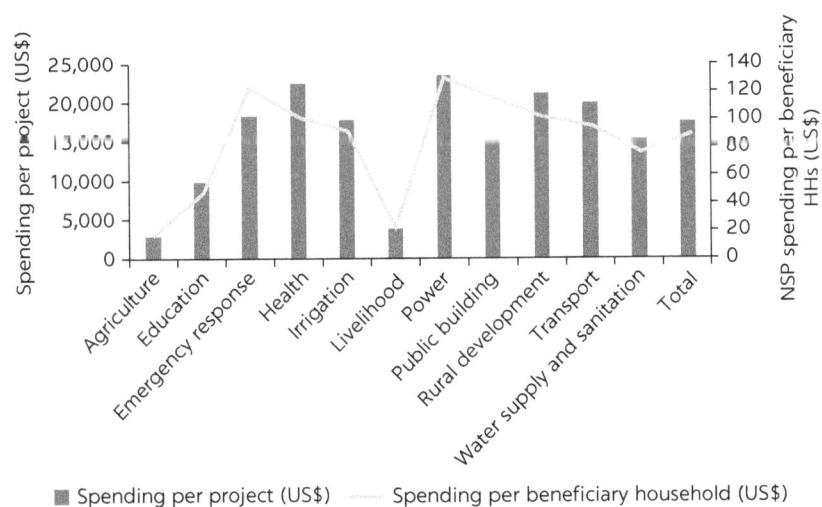

Spending per project (US$) — Spending per beneficiary household (US$)

Source: Project documents 2016.

FIGURE 4A.6

Intensity of NSP spending among beneficiaries, as of Sept. 2016

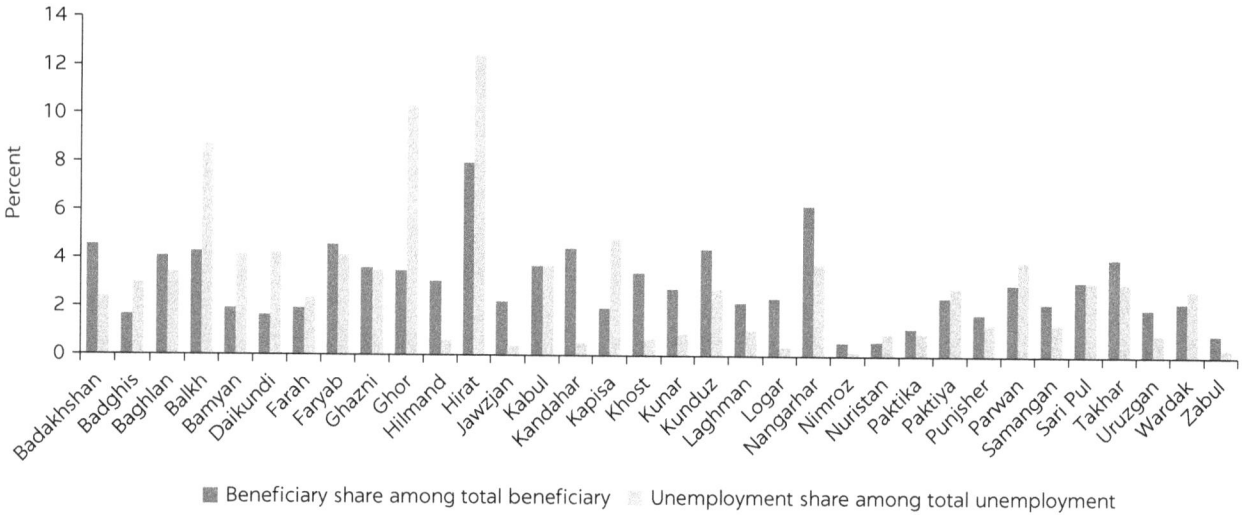

Source: Project documents 2016.

FIGURE 4A.7

Job creation (NSP), underemployment, and unemployment

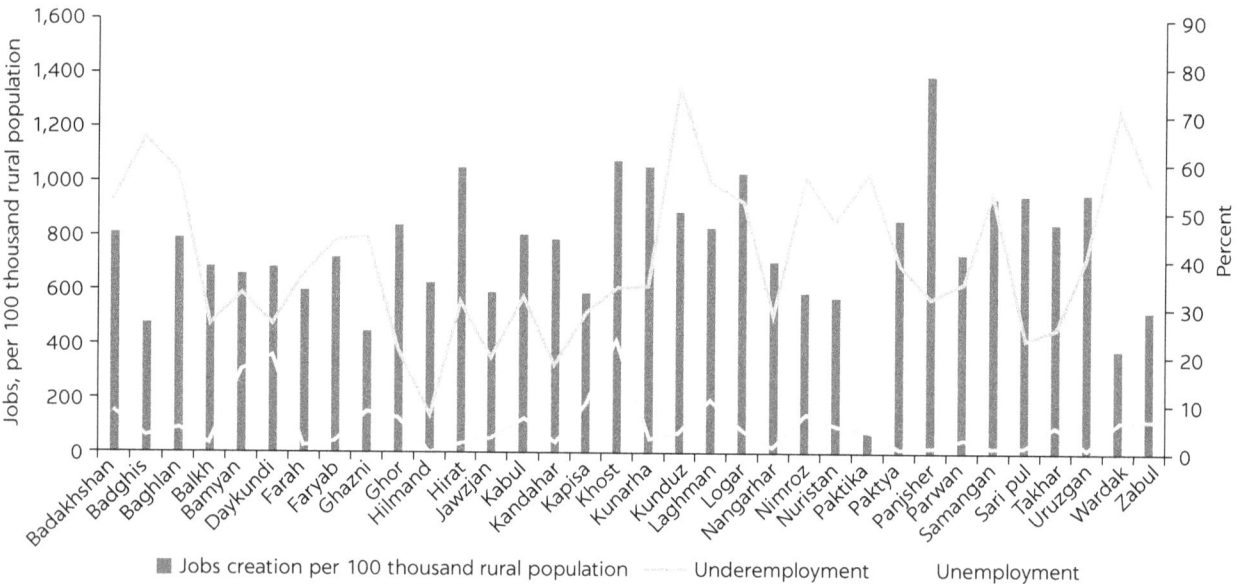

Source: Project documents 2016.

in Ghor province, unemployed people account for about 10 percent of Afghanistan's total rural unemployed, but beneficiaries account for about 3.5 percent of total project beneficiaries. Figure 4A.7 also shows that job creation resulting from NSP III spending is not always in provinces with severe unemployment and underemployment. While job creation through NSP activities may improve the employment situation in some provinces, we cannot establish causality due to a lack of data.

Figure 4A.8 presents landlessness, livestock market participation, and NSP spending across provinces. It shows some positive association between the severity of landlessness and the level of NSP spending. For example, landlessness is quite high in Nangarhar and Herat, as is NSP per capita spending. Livestock

FIGURE 4A.8

Intensity of NSP spending among beneficiaries, as of Sept. 2016

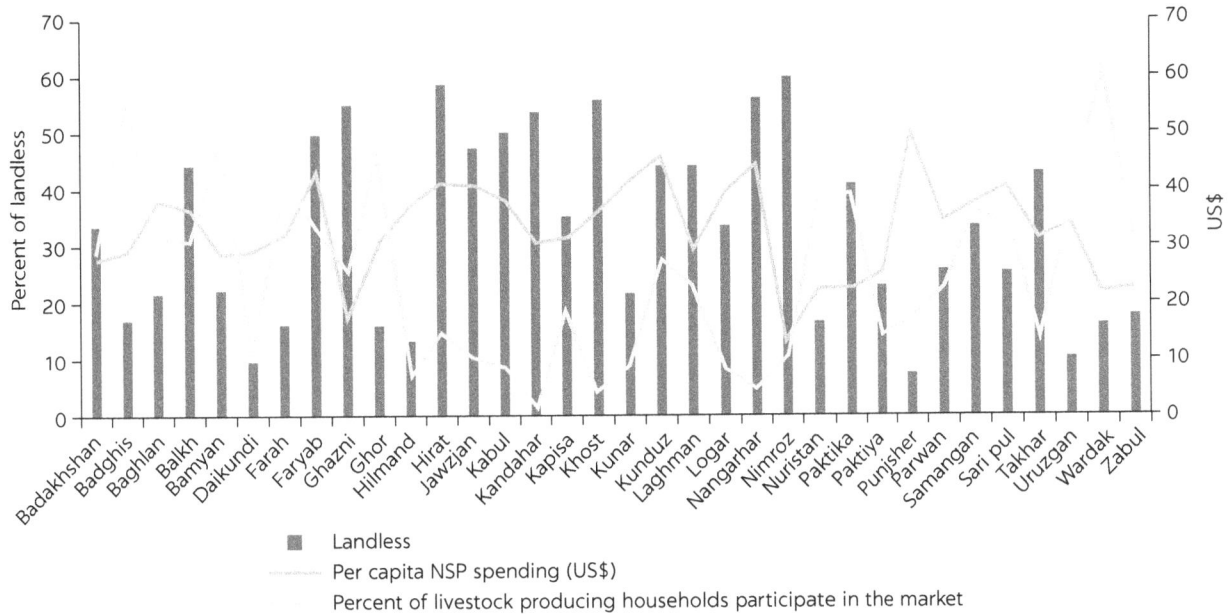

Landless
- - - Per capita NSP spending (US$)
Percent of livestock producing households participate in the market

Source: Project documents 2016.

BOX 4A.1

World Bank—Agriculture Global Practice core interventions in Afghanistan

Agricultural Production and Marketing

Delivery of extension services and linking smallholder farmers to intermediaries and markets, often through producer groups, are two key interventions under this area. The scope of extension comprises promoting agricultural research, developing extension materials for increasing farm productivity, and livestock and horticulture. To measure job outcomes under linking farmers to intermediaries and markets, the World Bank uses activities such as establishing producer groups that improve smallholders' access to marketing and technical advisory services, which improves the groups' organizational structure and increases their bargaining power. This process strengthens linkages between producers and traders, and provides access to regional, national, and/or export markets, which contributes to sustainable jobs.

Irrigation and On-Farm Water Management

Indicators used to analyze job outcomes include the increase in irrigated areas through construction or rehabilitation of irrigation schemes, or the number of

people trained water management-related practices. Evidence is well-documented on the improved levels of productivity, employment, and income resulting from these activities.

Rural Enterprise Development

Activities such providing rural entrepreneurs access to technical assistance, business development services, and financial products, and establishing producer organizations are used to analyze job outcomes. These activities result in better access to credit, working capital, and business services, all of which contribute to strengthening forward and backward linkages for rural enterprises and foster sustainable job creation.

Rural Livelihoods and CDD

Activities evaluated to analyze job outcomes include establishing local governance institutions that obtain grants to select, manage, and implement projects in their communities; capacity building of local government representatives; and subprojects to strengthen infrastructure, irrigation, and drainage.

market participation is also positively associated with NSP per capita spending in some provinces; however, we cannot draw any conclusions about the causal relationship due to a lack of relevant data.

NOTES

1. The NSP III project ended in 2017 and its successor, Citizens' Charter, has an even broader program.
2. The veterinary farm units under contract are involved in the Sanitary Mandate Contracting Scheme, the National Brucellosis Control Programme, and Farmer Field School extension activities. Their compound effect has been assessed as a 2 percent per-unit reduction in mortality in adult animals and a 3 percent reduction in calves. With an assumed value of Af 20,000 for such an animal at some stage, and using the current number of animals in the provinces where NHLP is active, it is possible to make a conservative estimate for the additional production/value that farmers reap from their animals.

   With better health care and nutrition, it is assumed that milk production has increased by 2 percent per 150-day lactation, which leads to a price of Af 30/liter to the incremental income stated in table 4.2. Similar conservative estimates have been made for sheep and goats. We estimate that each veterinary farm unit spends an additional four hours per week on the sanitary mandate, the brucellosis vaccination campaign, and the Farmer Field School activities. Multiplying this number by the number of contracted veterinary farm units in the program yields the number of additional paid-for working days the NHLP livestock project has generated. If we assume that farmers spend an additional two hours per day with their animals due to training and extra time for care, this will then yield job creation. The additional income could be considered as the income generated from this work; overall, it shows that farmers had an incremental income of $1.7 per working day in addition to the income that already existed.
3. http://aredp-mrrd.gov.af/2016.
4. Direct employment: A person (self-employed) or business/enterprise, group, association, etc. that occupies time and pays another person or organization to provide specific time- or deliverable-based services/products for a fixed period (full- or part-time job, contract, daily/hourly wage).

   Seasonal employment: Seasonal employees are hired to work on a part-time basis (daily wages, contract, per-item production) by SMEs and entrepreneurs that need extra help with increased work demand during a season. For example, seasonal employees in the agriculture sector may work in the handicrafts, tailoring, or embroidery subsectors. Seasonal employees work less than 120 days a year. These do not have to be consecutive days, and people may work for more than one employer at a time.

   Indirect employment: In the case of SMEs, one indirect employment is generated if the enterprise's income increases by Af 20,000 compared to their income last year. For micro-enterprises and economic groups, an increase of Af 10,000 is considered to generate one employment.

REFERENCES

CSO (Central Statistical Office). 2014. "Demographic and Social Statistics." Central Statistical Office, Kabul.

FAO and UNIDO. 2009. *Agro-Industries for Development U.K.* CAB International MPG Book Group.

Global Donor Platform for Rural Development, World Bank, and FAO (Food and Agriculture Organization of the United Nations). 2008. "Tracking Results in Agriculture and Rural Development in Less-Than-Ideal Conditions: A Sourcebook of Indicators for Monitoring and Evaluation." Global Donor Platform for Rural Development, World Bank, and FAO, Washington, DC.

IFC. 2013. "IFC Jobs Study: Assessing Private Sector Contributions to Job Creation and Poverty Reduction." IFC, Washington, DC.

ISR. 2016. The On-Farm Water Management Project. World Bank, Washington, DC.

MRRD (Ministry of Rural Rehabilitation and Development). 2016. Afghanistan Rural Enterprise Development Project. MRRD, Kabul.

World Bank. 2013. *World Development Report 2013: Jobs*. Washington, DC: World Bank.

——. 2017. "Implementation Status and Results Report. 2017." The National Solidarity Program III, World Bank, Washington, DC.

5 Summary and Policy Recommendations

Major findings and messages from the report are presented below, followed by policy recommendations.

SUMMARY MESSAGES

Afghanistan's rural labor market is grappling with high unemployment and underemployment rates, coupled with low absorption potential. The key factors defining the state of fragile labor markets in the agriculture sector are low agricultural income despite a high rate of agricultural employment, a lack of crop diversification, and poor linkages to markets for agriculture and livestock.

The labor market in rural Afghanistan is experiencing a new challenge: an influx of youth workers into the labor force that is creating stark competition for each new job. Moreover, this young generation is equipped with higher education and more competitive human capital. However, job creation has not kept pace with increased labor supply and youth have very high unemployment and underemployment rates, as well as a high unpaid labor force participation rate. More and more working-age people are joining the labor force, but the rural economy is unable to absorb them into the labor market, leaving many without paid employment or with underemployment. The key challenge for policymakers and development practitioners in Afghanistan is not only to generate more jobs, but better skilled and more inclusive jobs for youth and underemployed workers.

The vulnerable groups in rural areas, such as women, youth, and workers from the bottom 40 percent of income earners, are affected the most due to the fragile state of the labor market. Among the employed, more women and youth are unpaid family workers. The low level of market participation among women and youth could also be the reason for the high share of unpaid family workers among the women and youth. Without many paid opportunities in the nonfarm sector, young workers, even those with higher schooling, engage in their own households' agricultural activities as unpaid family workers. In addition to the lack of paid nonfarm opportunities, female workers face other challenges in finding nonagricultural work, such as access to finance and socio-cultural difficulties, and continue working as unpaid family workers.

Crop agriculture remains less diversified and overly concentrated in wheat production. Due to a near stagnant trend in wheat prices in recent years, crop agriculture has become less remunerative to farmers. This caused opium cultivation to gain momentum since 2010. Agricultural diversification toward high-value horticultural crops and livestock has not gained its potential momentum in rural areas. In addition, pervasive conflict has destroyed much of the local infrastructure that was key for fruit and livestock producers to access markets. Only a small number of rural households that own garden plots participate in the market and earn income from orchards. Similarly, the market participation level is much less among rural households that raise livestock.

Rural nonfarm activities are not moving much toward more large-scale manufacturing activities or formal sector employment. Construction and manufacturing are major sectors in terms of employment in rural areas. Yet, a large share of construction workers are day laborers and about 25–30 percent of manufacturing workers are unpaid family workers. Therefore, employment in these sectors may not be generating decent or sustainable long-term jobs, but engaging people on a more temporary basis.

Evidence from World Bank interventions in four projects suggests that the development of community-based enterprises, integrated value chains in rural areas, improved access to services and resources via NGOs and government agencies, improved technologies in livestock and orchards, and efficient water use can create more, sustainable, and inclusive jobs. While all four projects—the Afghanistan Rural Enterprise Development Program, the National Horticulture and Livestock Project, the National Solidarity Program III, and the On-Farm Water Management Project—play a crucial role supporting the improvement of rural livelihoods and generating and supporting jobs for rural people, the scale of some of their operations is insufficient to have a larger impact on the overall rural unemployment and underemployment situation.

A private sector case study revealed that supply chain network has significant potential for sustaining jobs. The case study assessed the distribution of jobs through backward and forward linkages of an established lead fruit processor. It found that, over the last five years, for every 10 jobs in the lead firm, the supply chain network has been supporting an average of approximately 8.5 jobs. Regardless, the political and security contexts in some regions present significant challenges to businesses and enterprises in the upstream and downstream of the agricultural value chains in rural areas.

POLICY RECOMMENDATIONS

Analyses of rural employment patterns and dynamics offer many insights about the state of Afghanistan's rural labor market and provide guidance for formulating effective job creation policies for the rural population. Based on the results of this study, we offer the following recommendations for supporting more, better, and inclusive jobs through agriculture and rural development.

Diversification toward high-value crops and livestock. Although policies to improve crop productivity, especially wheat productivity, should be in place, policies to diversify agriculture toward high-value agriculture (such as fruits, vegetables, and livestock) should be prioritized. Expansion of irrigation facilities and improved seeds availability can support productivity growth in crop agriculture and the reduction of underemployment among subsistence farmers. Bringing new areas under irrigation can generate more jobs in rural communities. Horticulture and livestock also have great potential for sustainable and inclusive job creation. Many rural households rear livestock and own orchard plots, but most are subsistence in nature and do not produce horticulture and livestock products commercially. Thus, policies and investments are needed to catalyze rural households that have potential to join commercial horticulture and livestock production.

Linking farmers to markets through continued investment in connectivity and rural infrastructure. The low level of market participation and highly subsistence nature of agriculture lead to a high number of unpaid family workers in agriculture. Strengthening agricultural value chains is key to increasing productivity and generating paid jobs in agriculture. Continued investments in rural roads, information and communication technology, reliable and affordable access to energy, and local infrastructure are necessary to enable local producers of crops and horticulture and livestock products to access markets and increase agricultural productivity. Rural infrastructure and improved rural-urban connectivity play a crucial role in the development of national value chains for agricultural products. Specific policies and investments to improve women's access to markets are also very important to catalyze the livestock, horticulture, and manufacturing and processing sectors where women are predominantly employed.

A balanced development strategy for an enabling environment for farm and nonfarm sectors. This is a priority for rural areas. There is strong evidence that rural sectors are interdependent, so both the farm and nonfarm sectors must be targeted for sustainable growth and employment generation. Increased agricultural productivity can boost demand for nonfarm services and products, and a vibrant nonfarm sector can increase demand for high-value agricultural products. Thus, the sectors support each other, raising productivity and generating more, sustainable, and inclusive jobs in rural areas. To operationalize this balanced development strategy, the scale of operations in the agriculture sector can be further developed to strengthen forward, backward, and consumption linkages, providing opportunities to establish value chains that, if exploited adequately, can support economic growth in the on-, off-, and nonfarm economies.

Access to finance and provisions for skills training for job creation in the nonfarm sector. These also need to be prioritized as key development strategies for rural people in Afghanistan, particularly women and youth. The analysis shows that literacy supports women to join the workforce. Evidence from agricultural and rural development interventions also shows that when women have access to finance and linkages to markets, they are successfully engaging in nonfarm activities and improving their livelihoods. Therefore, policymakers and donors need to stress policies and interventions that ease financial constraints and improve the skills of the rural workforce, mainly for the most vulnerable groups, to generate more, sustainable, and inclusive jobs.

Strengthening the private sector presence in agriculture and its linkage with the public sector: agribusiness. A strong private sector in agro-processing value chains, with public policy support, can spur job creation, improve the quality and productivity of existing jobs, and make jobs accessible to youth and women. The economy registers additional job creation through agriculture's backward and forward linkages with other sectors, generating jobs through indirect effects. Private sector efforts in agriculture should be adequately underpinned through macro institutional, regulatory, and business environment support to realize the sector's potential. This study shows that two policy levers can enhance the growth potential of jobs in the agro-processing sector. First, enhanced provisions of investments and advisory services to promising agro-processing firms are critical for strong job creation. This type of growth can create wage-bearing jobs for local economies, as well as in the regions from which inputs are sourced and where products are distributed and sold. Second, government policy must support the increased use of vertical integration to mitigate risks in the supply chain. Interventions to improve cross-sectoral linkages in the supply chain may offer agro-processing firms of all sizes better prospects to exploit market opportunities through flexible business models and lower capital requirements.

IMPROVING THE DESIGN STRUCTURE OF JOBS MEASUREMENT IN AGRICULTURE AND RURAL DEVELOPMENT

From the public-sector side, regarding "more jobs," it is important to articulate indirect jobs impacts and put more emphasis on measuring project benefits for the self-employed. For sustainable jobs, we recommend adjusting the ex-ante thought process before project design to increase the focus on how to improve labor productivity, strengthen forward and backward linkages between agriculture and agribusiness, and improve the earnings and social benefits of the jobs created. Inclusiveness needs to be visibly articulated for gender, as well as for other subgroups. Although gender has been successfully integrated into the design and results frameworks of most agriculture projects, the targeting of youth, the bottom 40 percent of income earners, and lagging regions needs to be improved and explicitly included in the project monitoring system.

To fully reap the agriculture sector's potential to create more, sustainable, and inclusive jobs, it is necessary to design and implement projects with a stronger and clearer jobs focus. To date, jobs results have often been mere by-products of development operations in agriculture and rural development. Explicitly considering the jobs challenge in the ex-ante project design and results framework will poise the agriculture portfolio to expand its impact on the multidimensional jobs agenda. Instead of a combination of complementary projects to achieve a sustainable impact, Afghan farmers need to have the necessary agricultural skills, marketing and trading knowledge, the required access to transport and markets, and a favorable macroeconomic environment so they can use available resources more effectively. This will also help create more sustainable and inclusive jobs at higher levels of value chains, which support higher employment intensity and inclusion.

From the private sector side, designing effective production and marketing strategies that help target job creation for educated workers and service providers throughout the agriculture value chains will require analytical rigor and advances in measuring the effects of proposed interventions. Policy planning in Afghanistan is challenging due to a lack of analytical evidence around the effectiveness of private sector interventions and their impact on employment and incomes. Improving the availability of administrative data and statistical assets can lower this information barrier and aid the design of interventions aimed at fostering job creation.

As the report noted, supplier and distribution linkages yield additional indirect employment growth throughout the economy. The extent of indirect job creation varies depending on the extent and nature (upstream or downstream) of inter-industry linkages. Ex-ante estimations of such effects are an important policy tool that allow policymakers to target sectors with higher indirect effects that are likely to be observed in other linked sectors. By knowing the extent of indirect job creation of each of the inter-industry linkages, we can target the sectors that will have higher indirect effects on employment. A common approach to model the impacts of interventions uses input-output matrices of the economy with industry-by-industry tables that describe inter-industry relationships in terms of products used in production processes. (Inter-industry relationships show the value of these products in response to increased demand in an industry.) These values can be translated into employment equivalents/effects using total employment data across all industries. An employment multiplier can be applied to the direct job creation effect to estimate the number of indirect jobs attributable to a specific intervention, a component of an economy-wide assessment of job creation impact.

Furthermore, wage outlays increase as businesses expand. When households can increase expenditures due to gainful employment, others in the economy benefit. Expanding economic activities leads to increased consumption of goods and services, termed as induced effects or income-type effects. Measuring these induced effects requires a lot of information, including data on employee incomes, household savings and expenditures, and geographical consumption patterns. The data need to be synthesized to compute increased revenues for businesses, which in turn are translated into the number of equivalent jobs that resulted from increased business revenues. Most of the data requirements are administrative in nature; national statistical agencies and affiliated government departments can make this data available.

Overall, an intensive policy discourse to create more, sustainable, and inclusive jobs in rural areas should be channeled through promoting farm, nonfarm, and off-farm linkages. The backward linkages with input suppliers (such as family farms, aggregators, and cooperatives) and service providers (such as transporters) further create jobs and income-bearing opportunities locally and in other regions. Similarly, through forward linkages with distributors, wholesalers, and retailers, agro-processors contribute to additional job creation and economic spillovers. Direct and indirect job creation is further complemented by income effects that result from rising incomes and expenditures on consumer goods and services, which require strong interlinkages between agricultural production and marketing activities in the public and private sectors.

www.ingramcontent.com/pod-product-compliance
Lightning Source LLC
Chambersburg PA
CBHW080426270326

41929CB00018B/3170